Climate Change Vulnerability and Communities in Agro-climatic Regions of West Bengal, India

"Professor Basu's book provides an insightful account of the effect of climate change on communities in West Bengal. The book will be valuable reading for both scholars and policy-makers".

Professor Kunal Sen, *University of Manchester, UK*

Jyotish Prakash Basu

Climate Change Vulnerability and Communities in Agro-climatic Regions of West Bengal, India

Theory and Practice

Springer

Jyotish Prakash Basu
West Bengal State University
Kolkata, West Bengal, India

ISBN 978-3-030-50470-0 ISBN 978-3-030-50468-7 (eBook)
https://doi.org/10.1007/978-3-030-50468-7

This Springer imprint is published by the registered company Springer Nature Switzerland AG
The registered company address is: Gewerbestrasse 11, 6330 Cham, Switzerland

Preface

The study of vulnerability due to climate change has drawn a lot of attention across the globe in general and India in particular. There are numerous studies available on the vulnerability analysis of the people in terms of poverty, famine, food insecurity, unemployment, inequality and deprivation. These studies are climate independent. India's economy is climate sensitive because most of the people are dependent on agriculture, forestry and fishery sectors which are climate dependent. The negative effects of climate change have multiplied the livelihoods risk of the marginalized people living in India. The most victims of climate change are smallholders, subsistence farmers, landless people and marginalized section of the people like women, casual labor and workers in the informal sector in India. There are few studies have focused on the impact of climate change on these vulnerable sections of communities in India. This study attempts to address the quantitative measurement of climate change vulnerability at the macro and micro level and identifies adaptation strategies of the households to cope with the adverse effect of climate change.

The study was conducted in the five different agro climatic regions West Bengal, India, namely hill region, foothill region, drought region and coastal regions during 2018–2019. 786 households and 17 villages in West Bengal are involved in the study. This research is associated with different sectors like agricultural sector, forestry sector, formal, informal sectors etc. In addition, the research measures climate change vulnerability of small holding farms, forest dependent communities, fishing and crab collecting communities, casual labourers, workers in the informal sectors and gender under different climatic conditions. This research has also identified the key vulnerabilities and the causes of such vulnerabilities and to what extent the remedial measures are taken into considerations. Different statistical and econometric models are applied to measure vulnerability and its determinants.

Particular attention is given to the role of the India's government associated policies like Sarva Shiksa Abhiyan, Mahatma Gandhi National Rural Employment Guarantee Act (MGNREGA), the housing scheme, Indira Awas Yojana, the Food for Work Programme, and the rural road building scheme, Pradhan Mantri Grameen Sadak Yojana etc. for vulnerability reduction measures.

Jyotish Prakash Basu
West Bengal State University
Kolkata
West Bengal
India

Acknowledgements

I would like to express my gratitude to the Indian Council of Social Science Research (ICSSR), New Delhi, for financial support to conduct the study. The study is based on secondary data and primary data. During the field survey, I received generous support, guidance and assistance from a number of people, including the students in different colleges and Universities, local people and many others in the selected nearby villages, as well as government officials in West Bengal. The field survey would not be possible without their active participations and cooperation. Special thanks in this regard should be given to the students and faculties of the Department of Economics of Sidho- Kanho- Birsha University, Purulia, students and faculties of the Department of Economics of St. Joseph's College, Darjeeling, and students and faculties of the Department of Economics of P.D. College, Jalpaiguri. Again thanks to Aishwarya Basu, PG student in Economics, Dept. of Economics, University of Calcutta, for her participation in the field survey.

I am further grateful to my University authority for giving me infrastructural facilities including space and library.

Jyotish Prakash Basu

Acknowledgements

[illegible]

[illegible]

Contents

Chapter 1
Introduction

1.1 Introduction

The study of vulnerability is a multidisciplinary approach. Different disciplines are involved in studying vulnerability research. It covers the areas of climate change as well as the other field of research like ecology, public health, poverty and development, secure livelihoods, sustainability science, and land—use change. In the development economics literature, vulnerability has its roots in famine, poverty, food-insecurity, unemployment, inequality, political economy and cultural practices etc. (Dercon and Krishnan 2000). Professor Sen (1984) defined vulnerability as deprivation which is the cause of failure to entitlements. This type of social vulnerability is climate independent.

In recent decades the world is experiencing a dramatic environmental and socio-economic changes due to population growth, rapid urbanization, poverty, environmental degradation and climate changes. The consequence of such changes yields larger size of vulnerable people in spite of anti-poverty measures are taken into consideration. Vulnerability to poverty addresses those households who are poor at present but they have a probability to become poor in the future and those who are non-poor at present but they have probability to be poor in future. Thus, poverty and vulnerability are the two sides of the same coin (Chaudhuri et al. 2002).

India is one of the most vulnerable countries in South Asia (Roy 2018). In India climate change is considered to be a major obstacle to development (Stern 2007), and likely to affect crop productivity adversely which leading to threatens food and livelihood security. In India 55% of its total working population are engaged in agriculture and contribute 14.1% of gross domestic product. Climate change projections report for India highlighted that an overall temperature would increase by 1–4 °C and precipitation by 9–16% towards 2050s (Kumar et al. 2011). Different agro-climatic regions are expected to exhibit different amount of rainfall, temperature and another climate related events are the increased frequency of occurrence of extreme events like droughts, floods, cyclones and earthquake and landslides.

J. P. Basu, *Climate Change Vulnerability and Communities in Agro-climatic Regions of West Bengal, India*, https://doi.org/10.1007/978-3-030-50468-7_1

India's economy is largely dependent on climate sensitive sectors such as agriculture, water resources and coastal zones, biodiversity and forestry (INCCA 2010). The severity of droughts, floods and storms has increased in various parts of India (NATCOM 2004). Annual mean surface air temperature is projected to increase between 3.5 to 4.3 °C by 2100 (MoEF 2012). These climate-induced impacts, along with changes in the frequency and intensity of extreme events such as floods and droughts, are likely to impact human health, food security, water resources, natural ecosystems, and biodiversity (MoEF 2012; Srinivasan 2012).

The negative impacts of climate change have multiplied the livelihoods risk of the marginalized people living in developing countries like India. The rural livelihoods in Indian Semi-Arid Regions (SARs) are predominantly dependent on natural resources, which are uncertain in nature. This uncertainty is a function of climatic risks (e.g. drought, extreme temperatures, erratic rainfall, high wind speed) (MoEF 2008) as well as dynamic non-climatic risks such as irregular employment opportunity, lack of infrastructure, inaccessibility of credit facilities, and erosion of social safety nets (Berrang-Ford et al. 2011; Deshingkar 2003; Agrawal 2010). The most victims of climate change are smallholders, subsistence farmers, landless people and marginalized section of the people like women, casual labour and workers in the informal sector.

In the context of climate change, vulnerability is the degree to which a system is unable to manage the harmful effects deriving from climate and environmental stress (Macchi et al. 2011). It depends on sensitivity, exposure and adaptive capacity (Adger 2006; Zahran et al. 2008). The progress of literature on vulnerability assessment in the last 10 years increased rapidly and is divided into five groups. First set of literature are related to the conceptual and methodological issues of climate risk, vulnerability and their assessment in general, as well as assessment implications for adaptation planning (Malone and Engle 2011; Funfgeld and McEvoy 2011; Hinkel 2011; Joakim et al. 2015; Dilling et al. 2015; Preston et al. 2011). The second group is concerned with the studies, utilizing a number of different indicators based methodologies, with visual representations of results (Rod et al. 2015; Binita et al. 2015; Wolf and McGregor 2013; Veerbeek and Husson 2013; Luers et al. 2003; Tonmoy et al. 2014) or a ranking of regions or countries (Brooks et al. 2005; Haddad 2005). The third group is associated with the measurement of vulnerability with the help of econometric based methodology (Hoddinott and Quisumbing 2003; Christiaensen and Subbarao 2004; Chaudhuri et al. 2002; Ligon and Schechter 2003). There are two other approaches in measuring vulnerability. The first one is sustainable livelihood framework of Hahn et al. (2009) consisting of different indicators which are used to analyze the vulnerability of sub-components and the overall vulnerability (Pandey and Jha 2012; Etwire et al. 2013; Shah et al. 2013; Aryal et al. 2013). Sustainable livelihood framework of Hahn et al. (2009) tries to integrate the interaction between the human being and its social and physical environment. The other is the Intergovernmental Panel on Climate Change (IPCC) framework approach which includes exposure, sensitivity and adaptive capacity as three major factors of vulnerability

(Sisay 2016). The fourth group of literature investigated the drivers and context of vulnerability (Morss et al. 2011; Luers 2005; O'Brien et al. 2007).

Fifth set of literature emphasized on the resilience building strategies for national and regional planning for reducing vulnerability (Brooks et al. 2005; Fussel 2007; Hinkel 2011). The objective of adaptation is the reduction of vulnerability and maintaining sustainable development (Cohen et al. 1998; Christie and Hanlon 2001; Markandya and Halsnes 2002; Swart et al. 2003; Agrawala 2004; Klein et al. 2007; Yohe et al. 2007). Addressing both poverty and vulnerability to climate change are two of the major challenges to sustainable development in the 21st century (TOI 2018).

One important dimension of vulnerability is the physical risks that result from climate stresses (Eriksen and O'Brien 2007). These stresses are likely to include increased temperature, sea-level rise, increased or decreased precipitation and changes in frequency and intensity of extreme events.

The entitlement and livelihoods literatures have focused on people's copying strategies in relation to climate variability (Sen 1981; Davies 1993; Glantz 1994; Scoones et al. 1996; Eakin 2006).

There is limited empirical work on vulnerability at household level though households are connected with larger socioeconomic process in the community and have great influence on decision making about climate change adaptation (Yaro 2006; Koirala 2015).

Given the above backdrop, the objectives of the study are the following;

First, is to construct a district level vulnerability indices in West Bengal and to examine factors affecting such vulnerability.

Second, is to measure the degree of vulnerability of the households in different agro-climatic regions of West Bengal and to examine the determinants of such vulnerability.

Third, is to formulate climate vulnerability indices for different occupational group of households say farmers, forest dependent community, tea garden workers, fishing communities, crab collecting communities, petty businessmen, casual labours, workers in the informal sector, formally employed etc. in different agro-climatic regions of West Bengal. In addition, the study also attempts to formulate gender wise vulnerability indices in different agro-climatic regions of West Bengal.

Fourth, is to identify possible adaptation options of the households to reduce vulnerability across agro-climatic regions in West Bengal.

Lastly, is to examine the role of Governmental policy to reduce climate related vulnerability.

1.2 Hypotheses

The following hypotheses are made

First, casual labourers, workers in the informal sectors, tea garden labourers, cultivators are more vulnerable to climate change than petty businessmen and workers in the formal sector.

Second, vulnerability is more acute in coastal areas than that of any other agro-climatic regions in West Bengal.

Third, household level vulnerability is influenced by non-climatic factors like poor infrastructure, inadequate credit facilities, off-farm income activities and non-farm income and other socio-economic variables (Agrawal 2010; Deressa et al. 2009).

References

Adger WN (2006) Vulnerability Global Environmental Change vol 16, 3rd edn, pp 268–281. https://doi.org/10.1016/j.gloenvcha.2006.02.006

Agrawal A (2010) Local institutions and adaptation to climate change. Soc Dimens Clim Change Equity Vulnerability Warm World, 173–198

Agrawala S (2004) Adaptation development assistance and planning: challenges and opportunities. IDS Bull 35:50–54

Aryal A, Brunton D, Raubenheimer D (2013) Impact of climate change on human–wildlife-ecosystem interactions in the Trans-Himalaya region of Nepal. Theor Alied Climatol 115:517–529. https://doi.org/10.1007/s00704-013-0902-4

Berrang-Ford L, Ford JD, Paterson J (2011) Are we adapting to climate change? Glob Environ Change 21(1):25–33. https://doi.org/10.1016/j.gloenvcha.2010.09.012

Brooks N, Adger WN, Kelly PM (2005) The determinants of vulnerability and adaptive capacity at the national level and the implications for adaptation. Glob Environ Change 15(2):151–163. Cambridge University Press Cambridge

Chaudhuri S, Jalan J, Suryahadi A (2002) Assessing household vulnerability to poverty from cross-sectional data: a methodology and estimates from Indonesia. Columbia University. Department of Economics. Discussion Paper Series 0102-52

Christiaensen L, Subbarao K (2004) Towards an understanding of household vulnerability in rural Kenya. J Afr Econ 14(4):520–558

Christie F, Hanlon J (2001) Mozambique and the great flood of 2000 James Currey. Oxford UK

Cohen S, Demeritt J, Robinson J, Rothman D (1998) Climate change and sustainable development: towards dialogue. Glob Environ Change 8:341–371

Davies S (1993) Are coping strategies a cop out? IDS Bull 24:60–72

Dercon S, Krishnan P (2000) Vulnerability seasonality and poverty in Ethiopia. J Develop Stud 36(6):25–53

Deressa TT, Hassan RM, Ringler C, Alemu T, Yesuf M (2009) Determinants of farmers' choice of adaptation methods to climate change in the Nile Basin of Ethiopia. Glob Environ Change 19:248–255

Deshingkar P (2003) Improved livelihoods in improved watersheds: can migration be mitigated? In: Watershed management challenges: improving productivity, resources and livelihoods International. Water Management Institute, Colombo. http://www.odi.org.uk/plag/RESOURCES/books/05_priya_migration_mitigation.pdf

Dilling L, Daly ME, Travis WR, Wilhelmi OV, Klein RA (2015) The dynamics of vulnerability: why adapting to climate variability will not always prepare us for climate change. Rev Clim Change 6:413–425

Eakin H (2006) Weathering risk in rural Mexico: climatic institutional and economic change. The University of Arizona Press, Tucson AZ

Eriksen SH, O'Brien K (2007) Vulnerability poverty and need for sustainable adaptation measures. Clim Policy 7(4):338–345

Etwire PM, Al-Hassan RM, Kuwornu JKM, Osei-Owusu Y (2013) Application of livelihood vulnerability index in assessing vulnerability to climate change and variability in Northern Ghana. J Environ Earth Sci 3:157–170

Funfgeld H, McEvoy D (2011) Framing climate change adaptation in policy and practice working Paper-1 VCCCAR Project: framing adaptation in the victorian context Melbourne Australia: Victorian Centre for Climate Change Adaptation Research

Fussel H (2007) Vulnerability: a generally applicable conceptual framework for CC research. Glob Environ Change 17:155–167

Glantz M (1994) Drought follows the plow: cultivating marginal areas. Cambridge University Press, Cambridge UK

Haddad BM (2005) Ranking the adaptive capacity of nations to climate change when Socio-Political goals are explicit. Glob Environ Change 15:165–176

Hahn MB, Riederer AM, Foster SO (2009) The livelihood vulnerability index: a pragmatic approach to assessing risks from climate variability and change—a case study in Mozambique. Glob Environ Change 19(1):74–88. https://doi.org/10.1016/j.gloenvcha.2008.11.002

Hinkel J (2011) Indicators of vulnerability and adaptive capacity: towards a clarification of the science–policy interface. Glob Environ Change 21:198–208

Hoddinott J, Quisumbing A (2003) Methods for micro econometric risk and vulnerability assessments social protection. Discussion Paper Series 0324. The World Bank Human Development Network, Washington DC

INCCA (2010) Climate change and India: A 4X4 Assessment Government of India New Delhi. http://www.natcomindia.org/natcomreport.htm

Joakim EP, Mortsch L, Oulahen G (2015) Using vulnerability and resilience concepts to advance climate change adaptation. Environ Hazards 14:137–155

KC Binita, Shepherd JM, Gaither CJ (2015) Climate change vulnerability assessment in Georgia. Appl Geogr 62–74

Klein RJT, Eriksen S, Næss LO, Hammill A, Tanner TM, Robledo C, O'Brien K (2007) Portfolio screening to support the mainstreaming of adaptation to climate change into development. Clim Change 84:23–44

Koirala S (2015) Livelihood Vulnerability Assessment to the Impacts of Socio-Environmental Stressors in Raksirang VDC of Makwanpur District Nepal. Master thesis Norwegian University of Life Sciences. The Department of International Environment and Development Studies Noragric www.nmbu.no

Kumar KK, Patwardhan SK, Kulkarni A, Kamala K, Koteswara Rao K, Jones R (2011) Simulated projections for summer monsoon climate over India by a high-resolution regional climate model (PRECIS). Curr Sci 101(3):312–326

Ligon E, Schechter L (2003) Measuring vulnerability. Econ J 113:C95–C102

Luers AL (2005) The surface of vulnerability: an analytical framework for examining environmental change. Glob Environ Change 15:214–223

Luers AL, Lobell DB, Sklar LS, Addams CL, Matson PA (2003) A method for quantifying vulnerability applied to the Yaqui Valley Mexico. Glob Environ Change 13:255–267

Macchi M (2011) Framework for community-based climate vulnerability and capacity assessment in mountain areas. Special Publication. International Centre for Integrated Mountain Development (ICIMOD), Kathmandu Nepal

Malone EL, Engle NL (2011) Evaluating regional vulnerability to climate change purposes and methods. Wiley Inter-discip Rev Clim Change 2:462–474

Markandya A, Halsnæs K (2002) Climate change and sustainable development: prospects for developing countries. Earthscan, London
Ministry of Forest and Environment (MoEF) (2008) National Action Plan on Climate Change (NAPCC). Prime Minister's Council on Climate Change Ministry of Forest and Environment Government of India. http://www.moef.nic.in/sites/default/files/Pg01-52_2.pdf
MoEF (Ministry of Environment and Forests) (2012) India's Second National Communications to the United Nations Framework Convention on Climate Change. Ministry of Environment and Forests New Delhi
Morss RE, Wilhelmi O, Meehl GA, Dilling L (2011) Improving societal outcomes of extreme weather in a changing climate: an integrated perspective. Annual Rev Environ Resour 36:1–25
NATCOM (2004) India's Initial National Communication to the United Nations Framework
O'Brien K, Eriksen S, Nygaard LP, Schjolden A (2007) Why different interpretations of vulnerability matter in climate change discourses. Clim Policy 7:73–88
Pandey MK, Jha A (2012) Widowhood and health of elderly in India: examining the role of economic factors using structural equation modeling. Int Rev Allied Econ 26(1):111–124. https://doi.org/10.1080/02692171.2011.587109
Preston BL, Yuen EJ, Westaway RM (2011) Putting vulnerability to climate change on the map: a review of approaches benefits and risks. Sustain Sci 6:177–202
Rod JK, Opach T, Neset TS (2015) Three core activities toward a relevant integrated vulnerability assessment: validate visualize and negotiate. J Risk Res 18:877–895
Roy AK (2018) Coastal communities and climate change: a study in Gujarat India environmental analysis and ecology studies vol 2, no 1
Scoones I, Chibudu C, Chikura S, Jeranyama P, Machaka D, Machanja W, Mavedzenge B, Mombeshora B, Mudhara M, Mudziwo C, Murimbarimba F, Zirereza B (1996) Hazard and opportunities: farming livelihoods in Dryland Africa–lessons from Zimbabwe. Zed Books and International Institute for Environment and Development, London
Sen AK (1981) Poverty and famines: an essay on entitlement and deprivation. Clarendon Press, Oxford UK
Sen AK (1984) Resources values and development. Oxford Blackwell 497
Shah KU, Dulal HB, Johnson C, Baptiste A et al (2013) Understanding livelihood vulnerability to climate change: Alying the livelihood vulnerability index in Trinidad and Tobago. Elsevier. Research gate https://www.researchgate.net/publication/236576858, http://dx.doi.org/10.1016/j.geoforum.2013.04.004
Sisay T (2016) Vulnerability of smallholder farmers to climate change at Dabat and West Belesa districts North Gondar Ethiopia. J Earth Sci Clim Change 7:8. https://doi.org/10.4172/2157-7617.1000365
Srinivasan J (2012) Impacts of climate change on India. In: Dubash NK (ed) Handbook of climate change and India: development politics and governance. Earthscan, pp 29–40
Stern N (2007) The economics of climate change: the stern review. Cambridge University Press, Cambridge UK
Swart R, Robinson J, Cohen S (2003) Climate change and sustainable development: expanding the options. Clim Policy 3(S1):S19–S40
TOI RSJ (2018) The economic impacts of climate change. Rev Environ Econ Policy 12(1):4–25
Tonmoy FN, El-Zein A, Hinkle J (2014) Assessment of vulnerability to climate change using Indicators: a meta analysis of the literature. Wiley Inter Discip Rev Clim Change 5:775–792. https://doi.org/10.1002/wcc.314 University of Oxford
Veerbeek W, Husson H (2013) Vulnerability to climate change: appraisal of a vulnerability assessment method in a policy context. K.F.C Report Number: 98 UNESCO-IHE OR/MST/177
Wolf T, McGregor G (2013) The development of a heat wave vulnerability index for London United Kingdom. Weather Clim Extremes 1:59–68
Yaro JA (2006) Is deagrarianisation real? A study of livelihood activities in rural northern Ghana. J Mod Afr Stud 44(1):125

Yohe GW, Lasco RD, Ahmad QK, Arnell NW, Cohen SJ, Hope C, Janetos AC, Perez RT (2007) Perspectives on climate change and sustainability In: Parry ML, Canziani OF, Palutik JP, van der Linden PJ, Hanson CE (eds) Climate change 2007: Impacts adaptation and vulnerability. Contribution of working group II to the Fourth Assessment Report of the Intergovernmental Panel on Climate Change. Cambridge University Press Cambridge, UK, pp 811–841

Zahran S, Brody SD, Vedlitz A, Grover H, Miller C (2008) Vulnerability and capacity: explaining local commitment to climate change policy in the United States. Environ Plan C Gov Policy 26:544–562

Chapter 2
Review of Literature

The present chapter reviews literature on vulnerability and adaptation to climate change. Literature on vulnerability is shown at the global, national and local level.

2.1 Literature on Vulnerability at the Global, National and Micro Level

There are some literatures focusing on micro level vulnerability of the rural households.

Piya (2012) analyzes the micro-level vulnerability of rural Chepang community in Nepal using indices based on exposure, sensitivity, and adaptive capacity. The indicators are weighted using Principal Component Analysis. The results of the study showed that the vulnerability of the poor people is due to low adaptive capacity.

Sisay (2016) analyzes climate change vulnerability of farm households in Dabat and West Belesa Districts of North Gondar, Ethiopia. Two Livelihood Vulnerability Index (LVI) approaches are used. One is composite vulnerability index and other is IPCC framework approach including exposure, sensitivity and adaptive capacity. The result reveals that farmers living in West Belesa are more vulnerable to climate change than farmers living in Dabat. The study has focussed on the improvement of education, the introduction of alternative means of livelihood, creation of access to market and road access, access to timely weather and early warning information etc.

Opiyo et al. (2014) uses statistical and econometric tools to measure households' vulnerability with the help of adaptive capacity, exposure and sensitivity in pastoral rangelands of Kenya. The study has used Principal component analysis (PCA) to select weights for different indicators. The study identified that 27% of households are more vulnerable, 44% of households are moderately vulnerable while 29% of households are less vulnerable to climate change. The study has used the ordered

J. P. Basu, *Climate Change Vulnerability and Communities in Agro-climatic Regions of West Bengal, India*, https://doi.org/10.1007/978-3-030-50468-7_2

Probit model to find out the determinants of vulnerability. The study has emphasised women's empowerment, education and income diversification for enhancing the resilience of the households.

Shah et al. (2013) measured and compared the vulnerability of two wetland communities say Nariva and Caroni in Trinidad and Tobago with the help of the Livelihood Vulnerability Index (LVI). The result of the study showed that the Nariva was more vulnerable than the Caroni. In terms of gender the female-headed households were more vulnerable than female-headed households.

Adu et al. (2017) applied Livelihood Vulnerability Index for measuring the vulnerability of small holding maize farming households in the Brong-Ahafo region of Ghana. The empirical results revealed that farming households in Wenchi municipality were more vulnerable than in Techiman municipality with respect to weather variability and adaptive capacity. The study calls for the initiation and implementation of climate change adaptation and household resilience projects by the government, donor agencies, and other related organizations in the two municipalities in the region.

Koirala (2015) estimate the level of livelihood vulnerability (Livelihood Vulnerability Index- LVI) for the Chepang community of Rakshirang Village Development Committee of Makawanpur district in Nepal and compare the vulnerability level between female-headed and male-headed households. The result shows that livelihood vulnerability as the product of poor infrastructures, limited access to basic public services such as education and healthcare, restricted access to the forest based natural resources and inadequate knowledge and skills on income generating activities.

Madhuri et al. (2014) studied the vulnerability of households in the seven blocks of Bhagalpur district in the state of Bihar. The study identified Naugachia as the least vulnerable block compared to the block Kharik because the former has better access to basic amenities and livelihood strategies than the latter.

Panthi et al. (2015) measured vulnerability for mixed agro-livestock smallholders in three districts around the Gandaki River Basin of central Nepal with the help of livelihood vulnerability index and IPCC vulnerability index. The result shows that both indices differed for mixed agro-livestock smallholders across the three districts, with Dhading scoring as the most vulnerable and Syangja the least.

Nkondze et al. (2013) investigated household vulnerability to climate change using household vulnerability index and the factors affecting vulnerability using multinomial logistic regression model of the households at Mpolonjeni Area Development Programme in Swaziland. The results show that 39.6% of the households were lowly vulnerable, 58.2% of were moderately vulnerable and 2.2% were highly vulnerable. The paper has focused on the health policy, rural development for the reduction of vulnerability.

Mudombi (2011) in his doctoral thesis used a multinomial logit model to find the factors influencing vulnerability. The identified factors are access to weather and socioeconomic factors like distance to the market and access to credit information etc.

Suzakhu et al. (2018) identified the factors which are responsible for vulnerability using logistic regression model. The factors are household head's education, irrigated land, non-agricultural income, and technologies that affect vulnerability of households, in farming communities at two locations in the Asian highlands.

There are some other literatures based on VEP focused on vulnerability to poverty at the household level in China (McCulloch and Alandrino 2003; Zhang and Wan 2006; Imai et al. 2010). These studies are based on five- year panel data of rural Sichuan households for 1991–1995 and concluded that demographic characteristics, education, the value of assets, and location are responsible for vulnerability.

Zhang and Wan (2006) analyzed vulnerability in six rural districts of Shanghai between 2000 and 2004 and concluded that low—education households are more vulnerable than that of high education households. On the basis of rural household survey data, Jalan and Ravallion (1998) found that a substantial fraction of poverty in rural China is transient in nature rather than chronic. Imai et al. (2011a, b) calculated various vulnerability measures in Viet Nam. Their findings showed that households in an ethnic minority group are more vulnerable than those in an ethnic majority group.

Based on expected poverty approach, Jha et al. (2010) have analyzed poverty and vulnerability in Tajikistan using a panel data set for 2004–2005 to describe the profile of vulnerable households and their findings showed that rural households are poorer and more vulnerable than urban households. Gaiha and Imai (2004) analyzed the vulnerability to poverty of rural households in South India during 1975–1984 using a variant of the expected poverty approach. Their findings showed that relatively rich households are highly vulnerable to long spells of poverty due to severe crop shocks occurred. Amin et al. (2003) analyzed vulnerability in Bangladesh using panel data incorporating idiosyncratic risks based on a risk- sharing test (Townsend 1994). Their conclusions on the basis of this test revealed that microcredit was successful at reaching the poor and unsuccessful at reaching the vulnerable poor.

Milcher (2010) utilized the expected poverty approach as a measure of vulnerability in Southeast Europe. He compared the profile of vulnerability to poverty for Roma and non-Roma households and found that Roma followed higher levels of vulnerability than non-Roma in terms of size of households, poorly educated head and involvement in informal activities.

The analysis of the effect of macroeconomic shock between 1985 and 1990 in Peru using a panel dataset by Glewwee and Hall (1998) showed that households headed by relatively well- educated persons, households headed by females, and households with fewer children are less vulnerable. In Papua New Guinea, Jha and Dang (2010) estimated poverty and vulnerability as expected poverty approach based on cross sectional data.

We have some other literatures focusing on the measurement of vulnerability in different geographical regions in the global and national levels. We may mention some studies in flood prone coastal regions (Huynh and Stringer 2018), Himalayan region (Koirala 2015; Panthi et al. 2015; Ives et al. 2000; Liu and Rasul 2007), African continent (Adu et al. 2017; IPCC 2007), Small Caribbean Island (Schneider 2007; McWilliams et al. 2005), and wet lands of south America (Shah et al. 2013).

2.2 Literature on Adaptation

Mohammed et al. (2014) identified the adaptation options and examined the factors on which adaptation depends under global climate change in coastal Bangladesh. Of 14 adaptation options irrigation ranks first while crop insurance ranks last. The factors identified for adaptation are age, education, family size, farm size, family income, and involvement in cooperatives. The paper also identified the constraints to coping strategies like lack of available water, shortage of cultivable land, and unpredictable climate.

To adopt need-based strategies to cope with nature's changing behavior, measurement of vulnerability is much urgent. Adaptation measure should be effective to fill up the gap between the need of the climate effected households if the vulnerability is assessed at the household level. The measures at the macro level fail to capture the problem at the micro level.

Although various methodologies for the measurement of vulnerability have been developed but the assessment methods have seen limited used in the context of India. In addition, there are limited numbers of studies available for the measurement of vulnerability due to climate change on small farmers especially in the context of India (Venkateswarlu and Singh 2015; Gopinath and Bhat 2012).

References

Adu DT, Kuwornu JKM, Anim-Somuah H, Sasaki N (2017) Application of livelihood vulnerability index in assessing smallholder maize farming households' vulnerability to climate change in Brong-Ahafo region of Ghana. Kasetsart J Soc Sci 2017:1–11

Amin S, Rai AS, Topa G (2003) Does microcredit reach the poor and vulnerable?: Evidence from Northern Bangladesh. J Dev Econ 70:59–82

Gaiha R, Imai K (2004) Vulnerability shocks and persistence of poverty: estimates for semi-arid rural South India. Oxford Dev Stud 32(2):261–281

Glewwe P, Hall G (1998) Are some groups more vulnerable to macroeconomic shocks than others? Hypothesis tests based on panel data from Peru. J Dev Econ 56:181–206

Gopinath M, Bhat ARS (2012). Impact of climate change on rainfed agriculture in India: a case study of Dharwad. Int J Environ Sci Dev 3:368

Huynh LTM, Stringer LC (2018) Multi-scale assessment of social vulnerability to climate change: an empirical study in coastal Vietnam. Clim Risk Manag 20:165–180

Imai K, Wang X, Kang W (2010) Poverty and vulnerability in rural China: effects of taxation. J Chin Econ Bus Stud 8(4):399–425

Imai K, Gaiha R, Kang W (2011a). Poverty dynamics and vulnerability in Vietnam. Appl Econ 43:3603–3618

Imai K, Gaiha R, Kang W (2011b) Poverty inequality and ethnic minorities in Vietnam. Int Rev Appl Econ 25:249–282

IPCC (2007) Climate change (2007). Impacts adaptation and vulnerability. Summary for policy makers. Intergovernmental Panel on Climate Change (IPCC). http://www.ipcc.cg/SPM.pdf

Ives JD, Messerli B, Spiess E (2000) Mountains of the world: a global priority land degradation and development 2. Parthenon Publishing Group, New York. https://doi.org/10.1002/(SICI)1099-145X(200003/04)11:2%3c197::AID-LDR390%3e3.0.CO;2-U

Jalan J, Ravallion M (1998) Transient poverty in postreform rural China. J Comp Econ 26:338–357
Jha R, Dang T (2010) Vulnerability to poverty in Papua New Guinea in 1996. Asian Econ J 24(3):235–251
Jha R, Dang T, Tashrifov Y (2010) Economic vulnerability and poverty in Tajikistan. Econ Change Restruct 43(2):95–112
Koirala S (2015) Livelihood vulnerability assessment to the impacts of socio-environmental stressors in Raksirang VDC of Makwanpur District Nepal. Master thesis Norwegian University of Life Sciences. The Department of International Environment and Development Studies. Noragric www.nmbu.no
Liu J, Rasul G (2007) Climate change the Himalayan mountains and ICIMOD. Sustainable mountain development, 5311–5314
Madhuri, Tewari HR, Bhowmick PK (2014) Livelihood vulnerability index analysis: an approach to study vulnerability in the context of Bihar. J Disaster Risk Stud 6(1):1–13
McCulloch N, Alandrino MC (2003) Vulnerability and chronic poverty in Rural Sichuan. World Dev 31(3):611–628
McWilliams JP, Cote IM, Gill JA, Sutherland WJ, Watkinson AR (2005) Accelerating impacts of temperature-induced coral bleaching in the Caribbean. Ecology 86(8):2055–2060
Milcher S (2010) Household vulnerability estimates of Roma in Southeast Europe. Camb J Econ 34:773–792
Mohammed NU, Bokelmann W, Entsminger JS (2014) Factors affecting farmers' adaptation strategies to environmental degradation and climate change effects: a farm level study in Bangladesh. Climate 2:223–241. https://doi.org/10.3390/cli2040223
Mudombi G (2011) Factors affecting perceptions and responsiveness to climate variability induced hazards. MSc Thesis University of Zimbabwe Faculty of Agriculture Department of Agricultural Economics and Extension Harare
Nkondze MS, Masku MB, Manyatsi A (2013) Factors affecting households vulnerability to climate change in Swaziland: a case of Mpolonjeni Area Development Programme (ADP). J Agric Sci 5(10)
Opiyo FE, Wasonga OV, Nyangito MM et al (2014) Measuring household vulnerability to climate-induced stresses in pastoral rangelands of Kenya: implications for resilience programming. Pastor Res Policy Pract 4(10). Pastoralism. Springer open Journal http://www.pastoralismjournal.com/content/4/1/10
Panthi J, Aryal S, Dahal P, Bhanddari P, Karakauer NY, Pandey VP (2015) Livelihood vulnerability approach to assessing climate change impacts on mixed agro-livestock smallholders around the Gandaki River Basin in Nepal. Reg Environ Change. https://doi.org/10.1007/s10113-015-0833-y
Piya L (2012) Vulnerability of rural households to climate change and extremes: analysis of Chepang households in the Mid-Hills of Nepal. PhD thesis Graduate School for International Development and Cooperation (IDEC) Hiroshima University Japan
Schneider SH (2007) In: Parry ML, Canziani OF, Palutikof JP, van der Linden PJ, Hanson CE (eds) Climate change 2007. Impacts adaptation and vulnerability. Contribution of working group II to the fourth assessment report of the intergovernmental panel on climate change. Cambridge University Press, pp 779–810
Shah KU, Dulal HB, Johnson C, Baptiste A et al (2013) Understanding livelihood vulnerability to climate change: applying the livelihood vulnerability index in Trinidad and Tobago. Researchgate https://www.researchgate.net/publication/236576858 Elsevier http://dx.doi.org/10.1016/j.geoforum.2013.04.004
Sisay T (2016) Vulnerability of smallholder farmers to climate change at Dabat and West Belesa Districts North Gondar Ethiopia. J Earth Sci Clim Change 7:8. https://doi.org/10.4172/2157-7617.1000365
Suzakhu NM, Ranjitkar S, Niraula RR, Salim MA, Nizami A, Vogt DS (2018) Determinants of livelihood vulnerability in farming communities in two sites in the Asian Highlands. Water Int 43(2):165–182 http://doi.org//10.1080/02508060.2017.1416445
Townsend R (1994) Risk and insurance in rural India. Econometrica 62(3):539–91

Venkateswarlu B, Singh AK (2015) Climate change adaptation and mitigation strategies in rainfed agriculture. In: Climate change modelling planning and policy for agriculture, Singh
Zhang Y, Wan W (2006) An empirical analysis of household vulnerability in Rural China. J Asia Pac Econ 11(2):196–212

Chapter 3
Data Base and Methodology

This chapter discusses data base and methodology used for the analysis of data. The data are collected both from secondary and primary sources. Different statistical and econometric techniques are used to analyze the data.

3.1 Secondary Data

The study is based on secondary data. Data like maximum temperature, minimum temperature, rainfall, food grains, no. of small farmers, cropping intensity, main agricultural population, net sown area, literacy rates, livestock, farm size, per capita income, length of the road, people below the poverty line and no. of people get medical facilities etc. are collected from Census report of 2001 and 2011 in 18 districts of West Bengal (Fig. 3.1).

3.2 Primary Data

Primary data are collected from five different agro-climatic regions of West Bengal.

3.2.1 Study Area and Sampling Design and Selection of Households

The primary data are collected from the purposively selected five agro-climatic regions of West Bengal. The selected agro-climatic regions are hill region of

J. P. Basu, *Climate Change Vulnerability and Communities in Agro-climatic Regions of West Bengal, India*, https://doi.org/10.1007/978-3-030-50468-7_3

Fig. 3.1 Location of different districts of West Bengal

Darjeeling district, foot hill region of Jalpaiguri district, drought region of :: Purulia district, coastal belt of Indian Sundarban in South 24 parganas district and East Midnapore district of West Bengal. These five districts are climate sensitive district of the state of West Bengal. Besides, three to four villages from each district are selected purposively. 10% households from each village are selected in probability proportionate to different livelihood groups as per their major occupation. Thus,

Table 3.1 Distribution of sample households across different agro-climatic regions of West Bengal

Agro-climatic Region	Village	No of sample households
Hill regions of Darjeeling district	Dupka Gaon	32
	Lamhatta	20
	Banekburn Tea Estate	25
	Manebhanjganj	73
	Sub total	150
Foot-hill regions of Jalpaiguri district	Mechbasti	36
	Detha Para	37
	Gomasta Para	57
	Sub total	130
Drought regions of Purulia district	Ajodhya	44
	Banduri	51
	Ebildi	22
	Matha	33
	Sub total	150
Coastal regions of Sunderbans in South 24 Parganas district	Bhagbatpur	30
	Laxminarayanpur	68
	Madhabnagar	51
	Paschim Dwarikapore	48
	Sub total	197
Coastal regions of Purba Medinipore district	Maitrapur	88
	Puba Mukundapur	71
	Sub total	159
	Total	786

total 786 households have been selected from 17 villages across five different agro-climatic regions of West Bengal. An interview method has been applied to collect data. In addition to the selected households, we discussed with the local people and interviews with local experts and school teachers and other knowledgeable elders in the villages. The selection of sample villages and sample households are presented in Table 3.1 as well as in the schematic diagram (Fig. 3.2) given (Fig. 3.3).

3.2.2 *Period of the Study*

The primary data are collected during the period of January 2018 to March 2019.

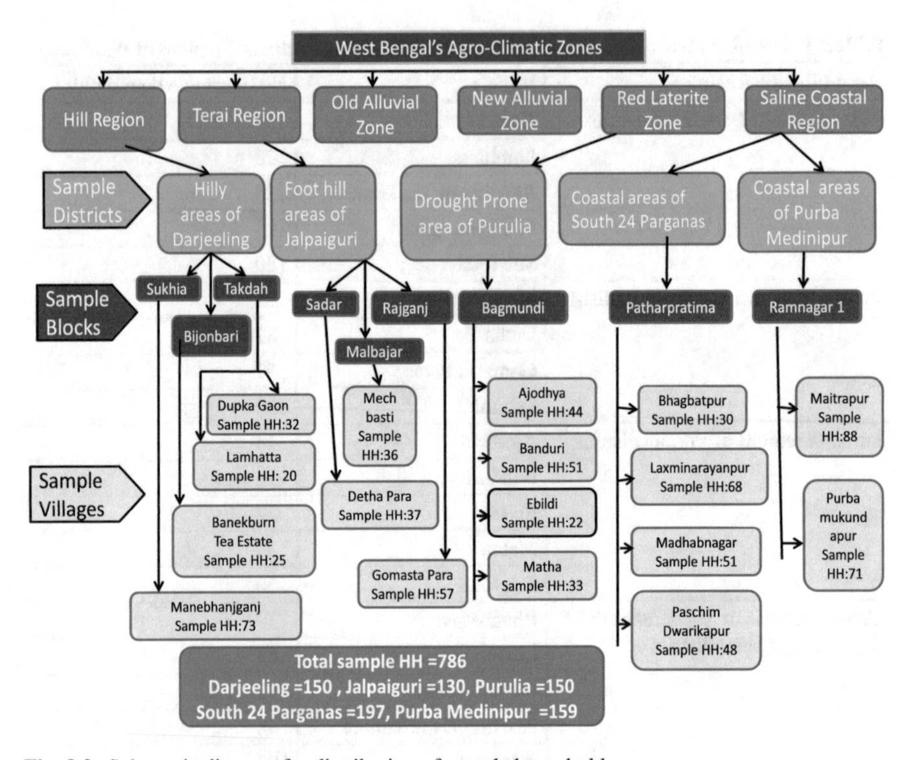

Fig. 3.2 Schematic diagram for distribution of sample households

3.3 Primary Data and Analytical Methods

The survey was conducted addressing the heads of each household with structured questionnaire containing socio-economic-demographic questions along with questions about realization of climate variables, local and administrative adaptive strategies. Socio economic data comprises age, caste, sex, occupation, education, landholdings, wealth, assets, income, dwelling status, drinking water, non-farm income, livestock assets, area under crops and consumption (food and non food consumption) etc. In addition, information on health related diseases, information on livelihood, food security, information on adaptation strategy, local knowledge, etc. formal and informal institutional support (formal extension, people to people extension, access to credit, social capital) have been collected (Figs. 3.4, 3.5, 3.6 and 3.7).

Analytical Methods: Vulnerability indices are used to measure vulnerability at the household level.

Vulnerability Index: In measuring vulnerability the present study follows two methods (a) Livelihood vulnerability Index (LVI) of Hahn et al. (2009) and (b) modified Livelihood vulnerability Index (LVI) following IPCC which is known as LVI_IPCC.

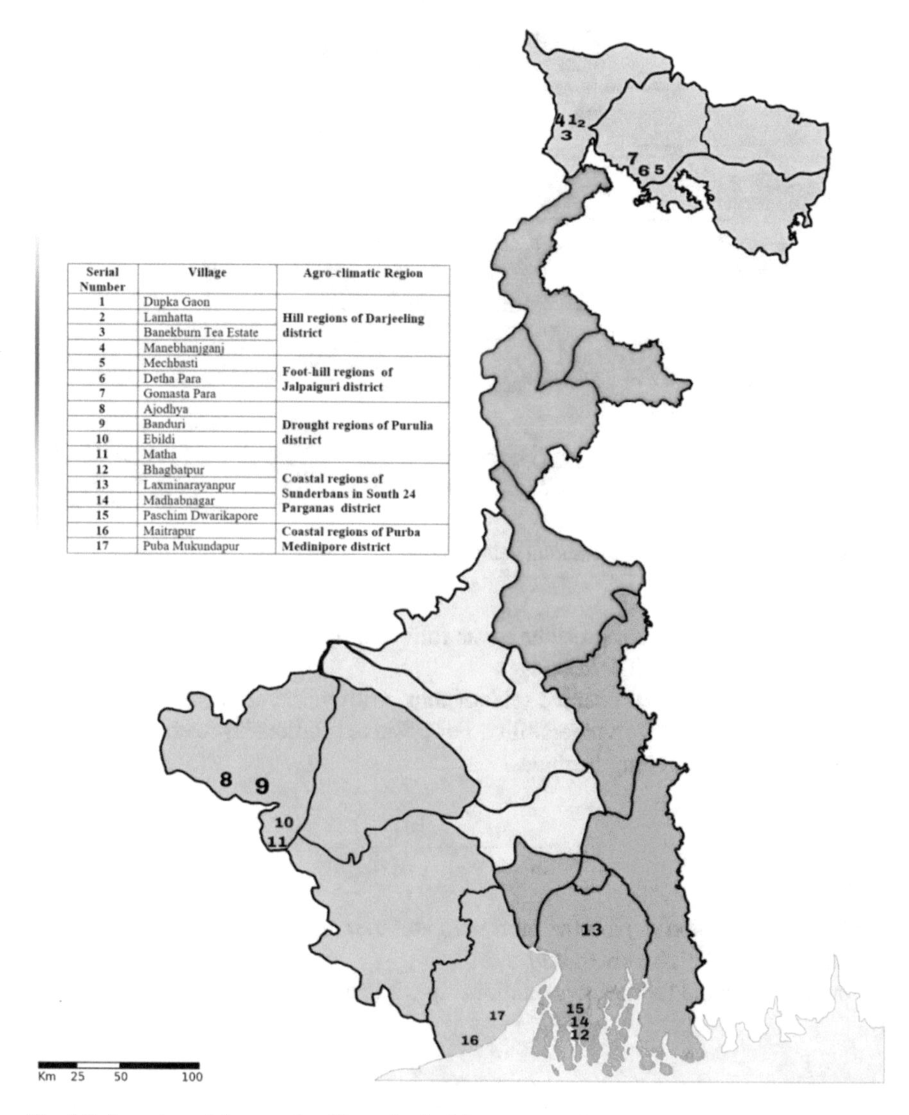

Serial Number	Village	Agro-climatic Region
1	Dupka Gaon	Hill regions of Darjeeling district
2	Lamhatta	
3	Banekburn Tea Estate	
4	Manebhanjganj	
5	Mechbasti	Foot-hill regions of Jalpaiguri district
6	Detha Para	
7	Gomasta Para	
8	Ajodhya	Drought regions of Purulia district
9	Banduri	
10	Ebildi	
11	Matha	
12	Bhagbatpur	Coastal regions of Sunderbans in South 24 Parganas district
13	Laxminarayanpur	
14	Madhabnagar	
15	Paschim Dwarikapore	
16	Maitrapur	Coastal regions of Purba Medinipore district
17	Puba Mukundapur	

Fig. 3.3 Location of the sample villages in the Map

In the calculation of livelihood vulnerability index (LVI) eight sub-components of vulnerability socio demographic profile (SDP), livelihood strategies (LS), food, social network (SN), natural capital (NC), water, health and climatic issues (C) are taken into consideration. Each of the sub-components index is again built with several indicators. The modified livelihood vulnerability index considers three contributory factors like exposure; sensitivity and adaptive capacity of vulnerability.

In order to formulate index we take normalize value of each indicator. The normalize value lies between 0 and 1. "0" shows minimum and "1" shows maximum

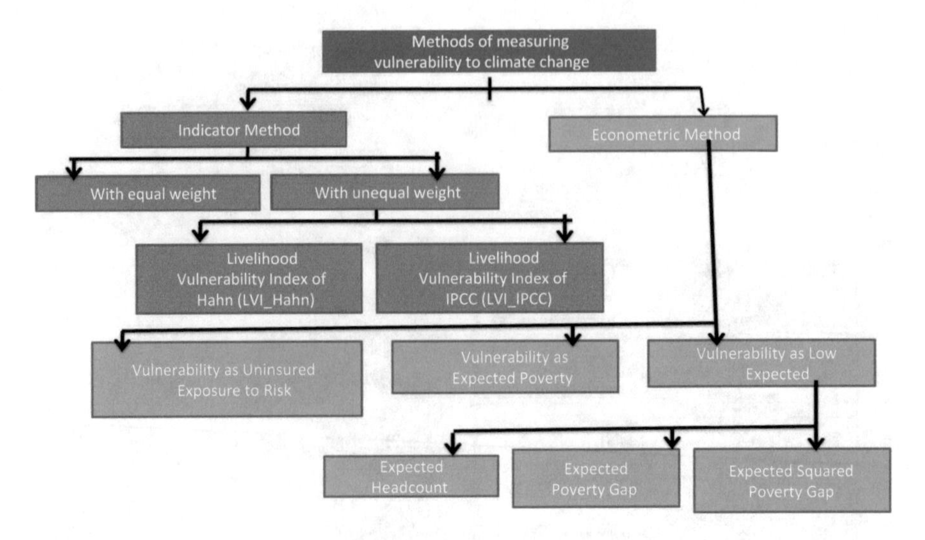

Fig. 3.4 Different methods of measuring vulnerability

values. This normalization procedure was followed by the methodology of Human Development Index (UNDP 2006).

Some indicators show positive relationships with vulnerability and others have negative relationship with vulnerability. For positive relationship with vulnerability we have used the following formula.

$$\mathrm{Y}_{ij} = \frac{y_{ij} - Min(y_{ij})}{Max(y_{ij}) - Min(y_{ij})} \tag{3.1}$$

Min(y_{ij}) and Max(y_{ij}) are the minimum and maximum values of the j indicator reflecting low and high vulnerability.

On the other hand for negative relationship with vulnerability, the formula is given by

$$\mathrm{Y}_{ij} = \frac{Max(y_{ij}) - y_{ij}}{Max(y_{ij}) - Min(y_{ij})} \tag{3.2}$$

The weighted vulnerability index is constituted by following method of (Iyenger and Sudarshan 1982).

$$\overline{W_i} = \frac{\sum_{j=1}^{k} k_j Y_{ij}}{\sum_{j=1}^{k} k_j} \tag{3.3}$$

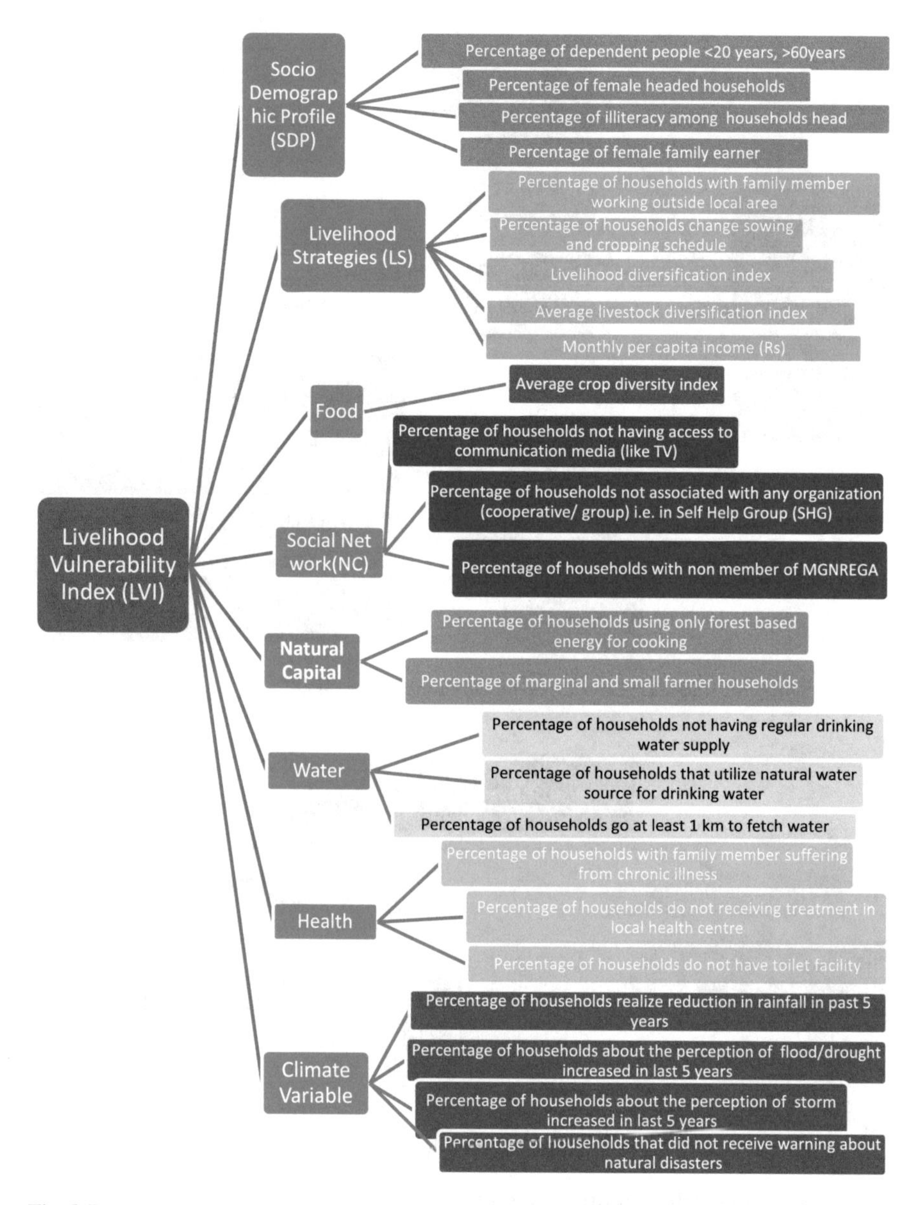

Fig. 3.5 Schematic diagram of livelihood vulnerability index (LVI_Hahn)

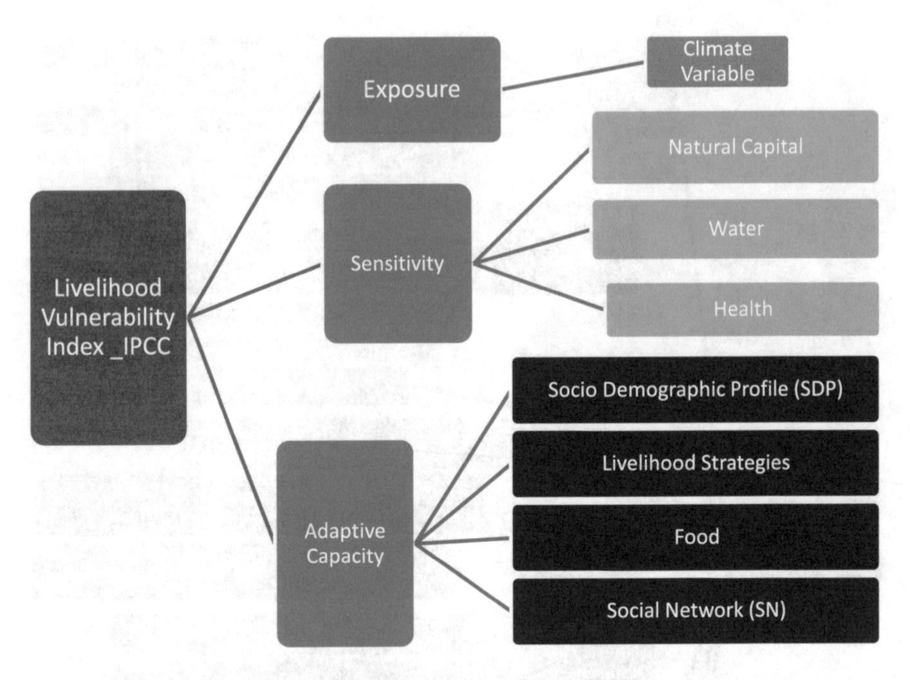

Fig. 3.6 Schematic diagram of livelihood vulnerability Index _IPCC

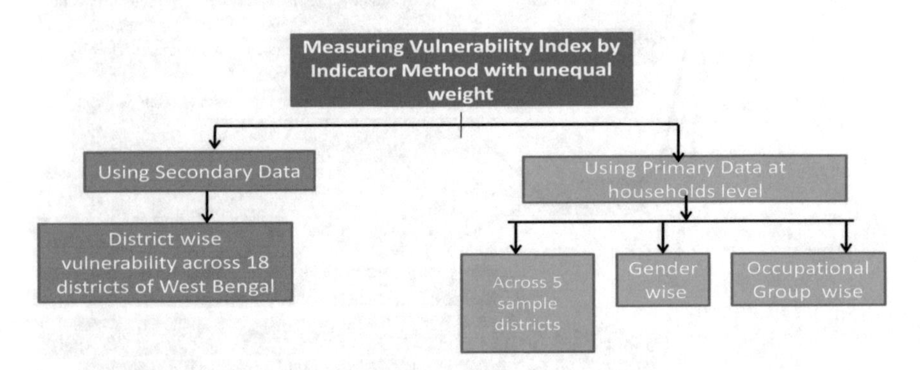

Fig. 3.7 Measurement of vulnerability

where $\overline{W_i}$ is the vulnerability Index of the ith subcomponent, Y_{ij} is the normalized score of jth indicator. k_j is the weights of jth indicators such that $(0 < k < 1 \; and \sum_{j=1}^{k} k_j = 1)$. $k_j y_{ij}$ is the weighted value of jth indicator of ith sub component.

Since $\sum_{j=1}^{n} k_j = 1$ Eq. (3.3) can be rewritten as

$$\overline{W_i} = \sum\nolimits_{j=1}^{n} k_j y_{ij}. \tag{3.4}$$

The weight is determined by

$$k_j = \frac{a}{\sqrt{var_i(y_{ij})}} \tag{3.5}$$

where 'a' is constant such that

$$a = \left[\sum_{j=1}^{j=n} \frac{1}{\sqrt{var_i(y_{ij})}}\right]^{-1} \tag{3.6}$$

After calculating 8 sub components, livelihood vulnerability index (LVI) is computed by weighted mean as follows:

$$\text{LVI} = \frac{\sum_{i=1}^{8} ki\overline{wi}}{\sum_{i=1}^{8} ki} \tag{3.7}$$

Since weight (ki) of ith sub component is 1 ($\because k_i = \sum_{j=1}^{n} k_j = 1$), Eq. (3.7) can be simplified as

$$\text{LVI} = \frac{\sum_{i=1}^{8} \overline{wi}}{8} \tag{3.8}$$

The livelihood vulnerability index varies from 0 (lowest value) to 1 (highest Value) (Pandey and Jha 2012).

LVI_IPCC index: The contributory factors of vulnerability are exposure, sensitivity and adaptive capacity. Adaptive capacity is constructed with four sub components (SDP), Livelihood Strategies (LS), and (SN).

$$\text{Adaptive capacity Index} = \frac{\sum_{i=1}^{4} ki\overline{Yi}}{\sum_{i=1}^{4} ki} = \frac{\sum_{i=1}^{4} \overline{wi}}{4} (\because ki = 1) = \frac{\overline{Ysdp} + \overline{Yls} + \overline{Yfood} + \overline{Ysn}}{4} \tag{3.9}$$

The sensitivity index is also based on three sub components like Natural Capital, Water and Health.

$$\text{Sensitivity Index} = \frac{\sum_{i=1}^{3} ki\overline{wi}}{\sum_{i=1}^{3} Wi} = \frac{\sum_{i=1}^{3} \overline{wi}}{3} (\because ki = 1) = \frac{\overline{Ync} + \overline{Ywater} + \overline{Ywater}}{3} \tag{3.10}$$

$$Exposure\ Index = Index\ of\ climate\ variable = \frac{\sum_{j=1}^{4} k_j y_j}{\sum_{j=1}^{n} k_j} \tag{3.11}$$

Lastly, LVI_IPCC index is then calculated as an average of exposure, sensitivity and adaptive capacity. Symbolically,

$$\text{LVI\_ IPCC} = \frac{Exposure + Sensitivity + AdaptiveCapacity}{3} \tag{3.12}$$

3.3.1 Panel Regression

After measuring vulnerability indices of 18 districts of West Bengal, the study now intends to identify factors of vulnerability based on secondary data. In order to estimate the determinants of vulnerability across the districts of West Bengal we apply panel regression model. We have checked the appropriateness of fixed effect model by Hausman test.

We have used fixed effect model the simplest version of the model is as follows:

$$\text{VI}_{it} = \text{c} + x'_{it}\beta + \varepsilon_{it}, \text{i} = 1, 2, 3, 4 \ldots\ldots 17; t = 1, 2 \tag{3.13}$$

where, i stand for the ith cross-sectional unit and t for the t-th time period. Here $t =$ 2001 and 2011.

The estimating equation is given by

$$\begin{aligned} \text{VI}_{\text{it}} = \text{c} + \beta_1\text{E}_{\text{i1}} + \beta_2\text{E}_{\text{i2}} + \beta_3\text{E}_{\text{i3}} + \beta_4\text{E}_{\text{i4}} + \beta_5\text{S}_{\text{i1}} + \beta_6\text{S}_{\text{i2}} + \beta_7\text{S}_{\text{i3}} + \beta_8\text{S}_{\text{i4}} + \\ \beta_9\text{S}_{\text{i5}} + \beta_{10}\text{S}_{\text{i6}} + \beta_{11}\text{A}_{\text{i1}} + \beta_{12}\text{A}_{\text{i2}} + \beta_{13}\text{A}_{\text{i3}} + \beta_{14}\text{A}_{\text{i4}} + \beta_{15}\text{A}_{\text{i5}} + \beta_{16}\text{A}_{\text{i6}} + \beta_{17}\text{A}_{\text{i7}} + \varepsilon_{\text{it}} \end{aligned}$$

where, The dependent variable VI_{it} is the vulnerability indices across the districts between 2001 and 2011

And 17 independent variables as described as

E1 District wise Average Max temperature (in °C)
E2 District wise Average Min temperature (in °C)
E3 District wise Average Rabi season rainfall (in mm)
E4 District wise Average Kharif season rainfall (in mm)
S1 District wise Population Density (Persons per km^2)
S2 District wise Production under Total Food grains (Production in thousand tonnes)
S3 District wise no. of Small farmers (in persons)
S4 District wise Cropping intensity
S5 District wise Main agricultural population (in person)
S6 District wise Net Sown area (% of geographical area)
A1 District wise Literacy rate (% of total population)
A2 District wise total no. live stock (per km^2)

A3 District wise % of average farm size
A4 District wise % of PCI (at constant price 2004–2005)
A5 District wise Length of Roads (in kilometre)
A6 District wise number of people live below poverty line (% of total population)
A7 District wise number of people get medical facilities (in person)
c Constant.

3.3.2 Econometric Model on Vulnerability to Expected Poverty

To start with Chaudhuri et al. (2002), vulnerability is measured by the given Eq. (3.14) as

$$V_{it} = \Pr\left(y_{i,t+1} = y\left(X_i, \beta_{t+1}, \varepsilon_{i,t+1}\right) \leq z | X_i, \beta_t, \alpha_i \varepsilon_{it}\right) \tag{3.14}$$

where, V_{it} is the probability of expected poverty of households i at time t. Where y is either income or consumption, z is treated as poverty line. X_i is the vector of household socioeconomic characteristics, β_t is a vector of parameters, α_i is a temporal household effect, and ε_i is the error term. Equation (3.1) shows the probability that a household will be poor in period $t + 1$ given the socio-economic condition prevails in period t.

Pritchett et al. (2000) extended Eq. (3.14) for more than one period and given by

$$R_i(n, z) = 1 - \left\{\left[\begin{matrix}\left(1 - \Pr\left(y_{i,t+1}\right) < Z\right), \ldots, \left(1 - \Pr\left(y_{i,t+n-1}\right) < Z\right), \\ \left(1 - \Pr\left(y_{i,t+n}\right) < Z\right)\end{matrix}\right]\right\} \tag{3.15}$$

where R_i (n, z) is the vulnerability or risk. Equation (3.16) implies the degree of vulnerability of household i is equal to 1 minus the probability of number of households to be poor.

Christiaensen and Subbarao (2004) defined vulnerability as the expected value of the Foster-Greer-Thorbecke (FGT) poverty measure, which is given as follows:

$$v_{i,y} = E\left[\left(\frac{z - c_i}{z}\right)^{\Upsilon}\right] = \int_0^z \left[\frac{z - c_i}{z}\right]^{\Upsilon} f(c_i)dc_i \tag{3.16}$$

Ligon and Schechter criticized the above Eq. (3.15) which does not consider the risk sensitivity. To remedy this drawback, Ligon and Schechter (2003) introduced expected utility approach which is known as defined vulnerability as low expected utility (VEU).

According to Ligon and Schechter (2003), vulnerability of household i is defined as

$$V_i = U_i\left(y^e\right) - EU_i(y_i) \tag{3.17}$$

where, U_i is a utility function which is positive and concave, y^e the level of income above which a household would be considered non vulnerable. Therefore y^e is to be chosen in such a way that inequality among individuals becomes zero.

Hence, Eq. (3.17) can be expressed as follows:

$$V_i = \left[U_i\left(y^e\right) - U_i(Ey_i)\right] + \left[U_i(Ey_i) - EU_i(y_i)\right] \tag{3.18}$$

where, the first term in the right hand side of the bracket shows utility gap (i.e. poverty). The second term shows the risk (shock) faced by household i. The second term is also decomposed into idiosyncratic and covariate risks, which is shown in Eq. (3.19). That is,

$$V_i = \left[U_i\left(y^e\right) - U(Ey_i)\right] + \left\{\left[EU_i\left(Ey_i|\bar{X}\right) - EU_i(y_i)\right]\right\} + \left\{U_i(Ey_i) - EU_i\left[E\left(y_i|\bar{X}\right)\right]\right\} \tag{3.19}$$

Equation (3.19) is further decomposed into measurement error and unexplained risk shown in Eq. (3.20).

$$\begin{aligned} V_i = &\left[U_i\left(y^e\right) - E(Ey_i)\right] + \left\{EU_i\left[\left(E(y_{it}|\bar{X}_t)\right] - EU_i\left[E\left(y_{it}|\bar{X}_t, X_{it}\right)\right]\right\} + \\ &\left\{U_i(Ey_{it}) - EU_i\left[E\left(y_{it}|\bar{X}_t\right)\right]\right\} + \left\{EU_i\left[E\left(y_{it}|\bar{X}_t, X_{it}\right)\right] - EU_i(y_{it})\right\} \end{aligned} \tag{3.20}$$

where

$[U_i(y^e) - E(Ey_i)] \Rightarrow$ Poverty
$\left\{EU_i\left[\left(E(y_{it}|\bar{X}_t)\right] - EU_i\left[E\left(y_{it}|\bar{X}_t, X_{it}\right)\right]\right\} \Rightarrow$ Idiosyncratic risk
$\left\{U_i(Ey_{it}) - EU_i\left[E\left(y_{it}|\bar{X}_t\right)\right]\right\} \Rightarrow$ Covariate risk
$\left\{EU_i\left[E\left(y_{it}|\bar{X}_t, X_{it}\right)\right] - EU_i(y_{it})\right\} \Rightarrow$ Measurement error and unexplained risk

The present study is based on cross section data followed by Chaudhuri et al. (2002). The expected consumption is estimated by using the following stochastic process

$$\ln C_h = X_h\beta + e_h \tag{3.21}$$

Here C_h is per capita consumption expenditure of h household, X_h shows household's socio-demographic-economic characteristics. "β" is represents the vector of parameters of each independent variable which is taken in the present model for analysis. e_h is the disturbance term which affects consumption expenditure of the households.

The variance of the disturbance term (e_h) is as follows:

$$\sigma_{e,h}^2 = X_h\theta \tag{3.22}$$

A three stage feasible generalized least squares (3FGLS) model (Amemiya 1977) is used to estimate the values of β and θ. At first, Eq. (3.21) is estimated by using ordinary least square (OLS) procedure. The estimated residual is used to estimate the following equation using ordinary least square (OLS) procedure:

The general form of the equation is the following;

$$\hat{\sigma}_{ols,h}^2 = X_h\theta + \eta_h \tag{3.23}$$

The estimation gives FGLS estimates, $\hat{\theta}_{\text{FGLS}}$. $\text{X}_h\hat{\theta}_{\text{FGLS}}$ is an efficient estimate of $\sigma_{e,h}^2$.

$$\sigma_{e,h}^2 = \sqrt{X_h\hat{\theta}_{\text{FGLS}}} \tag{3.24}$$

Using $\sigma_{e,h}^2$ Eq. (3.21) can be transformed to

$$\frac{lnC_h}{\hat{\sigma}_{e,h}} = \frac{X_h}{\hat{\sigma}_{e,h}}\beta + \frac{e_h}{\hat{\sigma}_{e,h}} \tag{3.25}$$

OLS estimation of Eq. (3.25) gives a consistent and asymptotically efficient estimate of β. These estimates $\hat{\beta}$ and $\hat{\theta}$ can directly be used to estimate expected log consumption:

$$\hat{E}(lnC_h|X_h) = X_h\hat{\beta} \tag{3.26}$$

and the variance of log consumption:

$$\hat{V}(lnC_h|X_h) = \hat{\sigma}_{e,h}^2 \tag{3.27}$$

With the help of the above Eqs. (3.21–3.27), the vulnerability of a households ($\hat{V}_h$) is given by

$$\hat{V}_h = \Pr\left(lnC_h < lnz|X_h\right) = \phi\left(\frac{ln\text{Z} - X_h\hat{\beta}}{\sqrt{X_h\hat{\theta}}}\right) \tag{3.28}$$

Where $ln\ z$ is the log of the minimum consumption level below which a household would be vulnerable. The study is based on the assumption that consumption per capita (or its log) follows normal distribution. The level of vulnerability depends on expected log per capita consumption expenditure ($\hat{\beta}$) and variance of log per capita consumption expenditure ($\hat{\theta}$). The level of vulnerability is inversely related with expected log per capita consumption while it is positively related with variance of log per capita consumption. In the present paper a sensitivity analysis has been done by taking three different level of minimum consumption. ϕ (.) denotes the cumulative density function of the standard normal.

3.3.3 Determinants of Vulnerability: Ordered Logit Model

The present study employs ordered Logistic model in order to identify the key variables that affect the household's vulnerability.

Following Greene (1997) the equation of ordered logistic model is given below:

$$Y_j^* = X_{ij}\beta + D_k + U_{ij} \tag{3.29}$$

Where Y_j is the dependent variable representing three category of outcome such as probability of less vulnerable: $Y = 1$, probability of moderate vulnerable: $Y = 2$ and probability of high vulnerable $= 3$. The X_{ij} are the explanatory variables. The study chooses explanatory variables like sex, age, household size, caste, education, marital status, nonfarm income, land holding, sanitation, and perception about drought, housing condition, adaptation and forest degradation. D_k is the agro-climatic region dummy. 'k' takes the values 1, 2, 3, 4 and 5 for corresponding to five sample agro-climatic regions such that D_1 represents hill region of Darjeeling district; D_2 represents foot-hill region of Jalpaiguri district; D_3 represents drought prone region of Purulia district; D_4 represents coastal region of Sundarban; D_5 represents coastal region of East Midnapore district. The disturbance term is denoted by U_{ij}.

3.3.4 Gender-Wise Vulnerability

For the measurement of vulnerability of the male headed and female headed households we have adopted Livelihood vulnerability Index (LVI) of Hahn et al. (2009) and the modified Livelihood vulnerability index which is known as LVI_IPCC.

After calculating the vulnerability indices for female and male headed households it is necessary to test whether there is a significant difference in the mean livelihood vulnerability indices along with major components for female and male headed households (Alhassan et al. 2018). To test this hypothesis, we employ t-statistic which is given below:

$$t = \frac{\mu_F - \mu_M}{\sqrt{\frac{\sigma_F^2}{N_F} + \frac{\sigma_M^2}{N_M}}}$$

Where μ_F *and* μ_M the mean values of vulnerability indices are for the female headed households respectively, while σ_F^2 *and* σ_M^2 denote the standard deviations of the vulnerability indices for the female and male headed households respectively. N_F *and* N_M are the sample sizes for female and male headed households respectively.

3.3.5 Probit Model for Adaptation Decisions

Probit model is used to determine the factors responsible for the decisions of adaptation to climate change. A Probit model is a type of regression where the dependent variable can take only two binary values, viz. 0 and 1. Y = 1 means adaptation takes place and "0" means no adaptation occurs.

Specifically, we assume that the model takes the form,

$$\Pr(\mathrm{Y} = 1 | \mathrm{X}) = \Phi(\mathrm{X}\beta)$$

Where Pr denotes probability and Φ is the Cumulative Distribution Function (CDF) of the standard normal distribution. The parameters β are typically estimated by maximum likelihood.

Suppose there exists an auxiliary random variable,

$$\mathrm{Y} = \mathrm{X}\beta + \varepsilon \quad (3.30)$$

Where $\varepsilon \sim N(0, 1)$. Then Y can be viewed as an indicator for whether this latent variable is positive:

$$\mathrm{Y}* = \begin{cases} 1, & \mathrm{Y} > 0 \\ 0, & otherwise \end{cases} = \begin{cases} 1, & -\varepsilon < \mathrm{X}\beta \\ 0, & otherwise \end{cases}$$

3.4 Terminology Involved in the Methodology

Livelihood diversification Index: Construction of livelihood diversification index has been done by taking score 1 for one livelihood activity, score 2 for two livelihood activities, score 3 for three livelihood activities, score 4 for four livelihood activities, score 5 for five or more livelihood activities. Hence the lowest value is 1 and the maximum value is 5.

Livestock diversification Index: Livestock diversification index was also constructed by taking score1 for household's holding 1 species in the herd of cattle, score 2 for household's holding 2 species in the herd of cattle, score 3 for household's holding 3 species in the herd of cattle, score 4 for household's holding 4 species in the herd of cattle, score 5 for household's holding 5 or more species in the herd of cattle.

Crop diversification index: household has the capacity to grow at least 1 additional crop such as vegetables or pulse along with traditional crop.

Indira Awaas Yojana: Indira Awaas Yojana (IAY) is a programme of Ministry of Rural Development, Government of India, to provide shelter to the homeless people. It caters to the needs of the poor people living below the poverty line (BPL) in Rural India.

References

Alhassan SI, Kuwornu JKM, Osei-Asare YB (2018) Gender dimension of vulnerability to climate change and variability: empirical evidence of smallholder farming households in Ghana. Int J Clim Change Strateg Manag 1756–8692. https://doi.org/10.1108/ijccsm10-2016-0156. Emerald Publishing Limited

Amemiya T (1977) The maximum likelihood estimator and the non-linear three stage least squares estimator in the general nonlinear simultaneous equation model. Econometrica 45:955–968

Chaudhuri S, Jalan J, Suryahadi A (2002) Assessing household vulnerability to poverty from cross-sectional data: a methodology and estimates from Indonesia. Columbia University. Department of Economics. Discussion Paper Series 0102-52

Christiaensen L, Subbarao K (2004) Towards an understanding of household vulnerability in rural Kenya. J Afr Econ 14(4):520–558

Greene WH (1997) Econometric analysis. New Jersey Prentice Hall

Hahn MB, Riederer AM, Foster SO (2009) The livelihood vulnerability index: a pragmatic approach to assessing risks from climate variability and change—a case study in Mozambique. Glob Environ Change 19(1):74–88. https://doi.org/10.1016/j.gloenvcha.2008.11.002

Iyengar NS, Sudarshan P (1982) A method of classifying regions from multivariate data. Econ Polit Weekly. Special Article. 2048–2052

Ligon E, Schecher L (2003) Evaluating different approaches to estimating vulnerability. Social Protection Discussion Paper Series No. 0410 Washington DC World Bank 2004

Pandey MK, Jha A (2012) Widowhood and health of elderly in India: Examining the role of economic factors using structural equation modelling. Int Rev Allied Econ 26(1):111–124. https://doi.org/10.1080/02692171.2011.587109

Pritchett L, Suryahadi A, Sudarto S (2000) Quantifying vulnerability to poverty-a proposed measure applied to Indonesia. Policy Research Working Paper Series 2437. Washington DC The World Bank

UNDP(2006) Human development report United Nations Development Program. http://hdr.undp.org/hdr2006/statistics/

Chapter 4
Profile of Study Area and Socio-Economic Analysis of the Sample Households

The present chapter discusses the profile of the study areas and the socio-economic conditions of the sample households. The study area comprises five different agro-climatic regions of West Bengal like hill regions of Darjeeling district, foot-hill regions of Jalpaiguri district, drought: regions of Purulia district, coastal Sunderbans of South 24 Parganas and another coastal region of East Midnapore district.

4.1 Profile of the Study Area

4.1.1 Hill Regions of Darjeeling District

The study area is the district of Darjeeling which is the part of eastern Himalaya. The district of Darjeeling is bounded by Sikkim in the north, Bhutan is in the northeast, Nepal lies in the west and Bangladesh is in the southeast. Tea garden, forestry and horticulture, tea industry and tourism are the major sources of livelihood of the people in the district. The economy of Darjeeling is dependent on tourism and tea cultivation. The major proportions of the people are of Lepchas, Limbus, Gorkhas, Sherpas and Bhutias. Total Literacy rate of the district is 79.6% which is above the state literacy rate. The Major proportion of the population consists of Lepchas, Limbus, Gorkhas, Sherpas and Bhutias. The annual mean maximum temperature is 14.8 °C and the minimum mean temperature is 6.9 °C. The average monthly mean temperature ranges between 8 and 18 °C. The rainfall data is showing increasing trend. The average annual rainfall was 298 cm during 1901–2001 and it rose to 316 cm during 2009–2013. Forest of Darjeeling district consists of sal, oak, semi-evergreen, temperate and alpine forests. This district faces water shortage which has been deterring factor for the expansion of economic activity and public health.

J. P. Basu, *Climate Change Vulnerability and Communities in Agro-climatic Regions of West Bengal, India*, https://doi.org/10.1007/978-3-030-50468-7_4

4.1.2 Foothill of Jalpaiguri District

The district Jalpaiguri is situated in the foot hill region of West Bengal. The district is surrounded by Assam hill in east and Darjeeling hill in the west and north-west. The district lies in the narrow stretch of land between Darjeeling Himalayas and Gangetic West Bengal. Geographically Jalpaiguri district is lies between 26° 15' 47" N–26° 59' 34" N latitudes and 88° 23'2"E–89° 7'30"' E longitudes. It covers 3386 sq km of geographical area. The boundary of the district covers two neighboring countries like Bangladesh in the South and Bhutan in the North. Jalpaiguri district is the gateway of Bhutan and northern states of India. In the west there is Darjeeling hill and Alipurduar and Coochbehar district in the west. A major part of the district is covered by forest consisting sal forest, deciduous forest. Though semi-moist—deciduous forest is the important portion of the forest but grassland in the valley of rivers provides fodder to the wildlife. Average annual rain fall is 3160 mm which occurs mostly due to south west monsoon. The month of January is the coldest with average temperature 11 °C which record lowest as 2 °C and May is the hottest month of the district with average maximum temperature is nearly 32 °C and it reaches maximum to 40 °C. Average annual humidity in the district is 82%. According to 2011 census total population of this district is 3872846 with sex-ratio 954 females for every 1000 males. In Jalpaiguri district prevalence of SC and ST population is very high. The literacy rate of the district is 79.79%. The primary sources of livelihoods of the people are agriculture with principal crops like Jute and paddy. Paddy is produced twice a year. Major section of the population is dependent on forest and tea gardens.

4.1.3 Drought Region of Purulia District

Geographically the district of Purulia lies between 22.60°N–23.50°N latitude and 85.75°E–0.22'W Longitude. The district covers 6259 square kilometer area. The boundary of the district covers the district of Bankura and West Midnapure in the east, the district of Burdwan and Dhanbad district of Jharkhand in the North, Bokaro and Ranchi districts of Jharkhand in the West, Singhbhum district of Jharkhand State. Drought prone Purulia district is a part of Jangle Mahal, poverty ridden (poverty rate 32.7%) and tribal dominated district. Agriculture is the main source of livelihood of the district. The main agricultural crops grown in this district are rice, maize, potatoes, groundnuts, wheat, etc. The climate of this district is sub tropical in nature. The district is known as rainfall deficient and high temperate district. The average rainfall of this district was 1189.2 mm in 2015–16 due to south west monsoon. The soil condition is red laterite type. The district experiences very high temperatures especially during summer. Maximum temperature is observed as 43.67 during summer in the month of April and May. The average temperature of Purulia is 25.41° centigrade. Minimum temperature is observed as 9 °C during winter in the month of December. Relative humidity is high in rainy season ranging from 75 to 85%

which come down to 25 to 35% in summer. Main workers are one fourth of the total work-force, marginal workers are also one fourth of the total labor force, and the rest 55.54% belongs to non work force. Among the main working class, Agricultural workers are of 36.09% specially followed by cultivators (31.24%), other workers are of 25.25% and rest of the workers are of 7.42% in household industry. In the district of Purulia, agriculture is predominantly mono-cropped. The forest area of this district is 29.69% of total geographical area.

4.1.4 Coastal Region of Sunderbans of South 24 Parganas District

Sundarban Delta is one of the Asian Mega Deltas with the highest population density and identified as the most vulnerable region (Nicholls et al. 2007). This delta was declared as World Heritage and Biosphere Reserve in 1989 by the *United Nations Educational, Scientific and Cultural Organization* (UNESCO) and the government of India respectively. This region is characterized by tropical cyclones, storm surges, land subsidence, sea level rise, coastal erosion and coastal inundation (Dey et al. 2016). According to Hajra et al. (2017) coastal belt of Indian Sundarban is frequently prone to climate hazard and most vulnerable region. On the other hand the Sundarban region is one of the richest ecosystems regions in the world. It is the largest tidal mangrove forests in Asia. Sundari and Gewa are the major trees grown in the Mangrove forest. From the name of tree Sundari, the delta is named as Sundarban (Paramanik 2014). Sundarban, part of Indian geographical territory, is located in between 21° 32'–21° 55' N latitude and between 88° 42'–89° 04' E longitude (www.wikipedia.org). It lies on the delta of Ganges, Bramhaputra and Meghna River.

In South 24 Parganas, the annual average maximum temperature is 35 °C and the minimum temperature is 18.5 °C during 2010–12. This area receives heavy rainfall and has an average humidity about 82% due to the proximity of the Bay of Bengal. Sundarban is rich in bio-diversity. Population of Sundarban of the district of South 24 parganas is 3.3 million (Census of India 2011). About 50% of the farmers are landless and large percentage of rural households is marginal farmers. Crops are mainly used for self consumption. The main sources of livelihood of the people of Sundarban are fishing, crab collection and honey collection with high exposure, sensitivity and low adaptive capacity (Hajra and Ghosh 2018; Danda 2007). They are adversely affected by increases in salinity due to sea level rise, intensity of storm, cyclones, coastal inundation and land erosions (WWF 2010; Lwasa 2014).

4.1.5 Coastal Region of East Midnapore District

East Midnapore district is located in the southern part of West Bengal which is surrounded by the river of Rupnarayan and Hooghly in the East. Latitude and longitude of the district are 21.9373°N and 87.7763°E respectively. It is a part of the lower Indo-Gangetic plain and Eastern coastal plain. This district experiences the highest literacy (87.66%) among the 18 districts of West Bengal as per Census 2011. The coastal region of East Midnapore attracts a large number of tourists throughout the year. The important tourist destination places of this district are Digha, Mandermani, Khathi, Tajpur, Sankarpur and Udaypur etc. The sources of livelihood of the people are cultivation, fishing and livestock rearing. People of East Midnapore district are engaged in producing different types of handicrafts like mat and mat products, cane and bamboo products, pottery and terracotta, sea shell etc. The climate of East Midnapore is tropical. Yearly average temperature is 26.8 °C and rainfall is 1436 mm. South west monsoon brings major portion of rain during June to August.

4.2 Description of Sample Households Across Five Agro-Climatic Regions

4.2.1 Description of Sample Households in the Hill Region of Darjeeling District

In this section we describe socio-economic-demographic features, infrastructural facilities, assets structure, livelihoods, income and consumption of sample households in the hill regions of Darjeeling district.

4.2.1.1 Socio-Economic and Demographic Features of Sample Households of Darjeeling District

The primary data was collected from 150 households across four villages in the district of Darjeeling. The key socio-economic and demographic features of sample households are shown in Table 4.1.

Social status: Majority of households (72.67%) belong to the scheduled tribe (ST) who are most disadvantageous section of the society. The ST population is highest in the village Manebhanjganj (84.2%) followed by the village Dupka Gaon (71.88%), Lamhatta (65%) and Banekburn Tea Estate (40%).

Gender: About 50% of households are female headed.

Age distribution: Majority of the households' head are middle aged between 41and 60 years.

Table 4.1 Socio-economic conditions of the households in Darjeeling District

Socio-economic variables	Banekburn Tea Estate (No. of HH = 25)	Dupaka Gaon (No. of HH = 32)	Lamhatta (No. of HH = 20)	Manebhanjgang (No. of HH = 73)	All (No. of HH = 150)
Social status					
ST	10 (40)	23 (71.88)	13 (65)	63 (84.2.1.3)	109 (72.67)
General	8 (32)	6 (18.75)	2 (10)	2 (2.74)	18 (12)
OBC	7 (28)	3 (9.38)	5 (25)	8 (10.96)	23 (15.33)
Gender					
Female	13 (52)	15 (44.2.1.88)	8 (40)	39 (53.42)	75 (50)
Male	12 (48)	17 (53.13)	12 (60)	34 (44.2.1.58)	75 (50)
Age of head of households					
21–40 years	3 (12)	11 (34.38)	5 (25)	33 (45.21)	52 (34.67)
41–60 years	14 (56)	14 (43.75)	6 (30)	31 (42.47)	65 (43.33)
above 60 years	8 (32)	7 (21.88)	9 (45)	9 (12.33)	33 (22)
Education					
Illiterate	4 (16)	5 (15.63)	9 (45)	14 (19.18)	32 (21.33)
Primary	2 (8)	3 (9.38)	3 (15)	16 (21.92)	24 (16)
Secondary	10 (40)	13 (40.63)	6 (30)	30 (41.1)	59 (39.33)
Above secondary	9 (36)	11 (34.38)	2 (10)	13 (17.81)	35 (23.33)
Average of family size	4.76	3.81	4.50	4.37	4.33
Economic status					
BPL	10 (40)	16 (50)	15 (75)	42 (57.53)	83 (55.33)
APL	15 (60)	16 (50)	5 (25)	31 (42.47)	67 (44.67)
Land holding (acre)					
Land less	21 (84)	32 (100)	10 (50)	72 (98.63)	135 (90)
<1 Acre	4 (16)	0 (0)	10 (50)	1 (1.37)	15 (10)

Source Field Survey, 2018; *Note* Figures in the parentheses represent percentage

Educational status: It shows that 21.33% of households head are illiterate which is highest in Lamhatta (45%) followed by Manebhanjganj, Banekburn Tea Estate and Dupka Gaon. About 63% households head possess either secondary level education or higher. This implies that basic education of Darjeeling district is strong.

Family size: Average number of family members in each household is 4.33. Only in Dupka Gaon household size is below 4.

Economic status: The economic status of the households is determined by the possession of the ration card issued by the government of West Bengal and landholdings. It is observed that more than half (55.33%) of households are living below the poverty line (BPL) and the rest 44.67% households are above poverty line. Highest percentage (57.53%) of BPL population is found in Lamhatta, followed by Manebhanjganj, Dupka Gaon and Banekburn Tea Estate. It is noted during the field visit though BPL population is 55.33%, but all households got subsidized rice (Rs. 2/Kg) and wheat (Rs. 2/Kg) in Khadya sathi prakalpa of Government of West Bengal.

Landholding of the households indicates that 90% of households do not have any cultivable land. All households are landless in Dupka Gaon.

4.2.1.2 Infrastructural Facilities of Sample Households in Darjeeling District

The infrastructural facilities like drinking water facility, sanitation, public health care facility, banking facility, housing etc. of sample households in Darjeeling district are displayed in Table 4.2.

Drinking water: It shows that spring water from hill is the prime source of drinking water followed by river water. Lamhatta and Dupka Gaon mainly depend on spring for drinking water whereas majority of households of Manebhanjganj are dependent on water through pipe line. **Sanitation facility**: The study finds that 88.67% of households have sanitation facility. Most (64%) of the sanitary construction are normal and rest (23.33%) are Pucca.

Public health care facility: 77% of households access public health care facility. It is highest in Manebhanjganj (95.89%) and lowest in Lamhatta (35%).

Banking facility: More than 80% households have the banking facility. It is the major indicator of financial inclusion which is signifies the wellbeing of the households.

Sources of credit: The sources of credit are banks and money lenders. The study finds that about 24.67% of households taking loans from banks and 20.67% of households are taking loan from money lenders.

Housing condition: More than 60% household are living in the pucca houses, 34.67% households are living in Kachha houses while rest 4.67% people are living in semi-pucca houses.

4.2.1.3 Household's Assets in Darjeeling District

Two types of assets like livestock and physical assets are taken into the analysis.

Livestock asset: Livestock asset comprises cow, hen, pig and goat etc. The average number of hen per household owned is highest (8.51) followed by pig (2.61), Goat (0.07) and cow (0.41) (Table 4.3). Average number of hen is highest in Dupka Gaon and lowest in Manebhanjganj. No household in Manebhanjganj villages possess goat. In rearing of pig Manebhanjganj is on the top while Dupka Gaon Tea Estate

Table 4.2 Infrastructural facilities in Hill region of Darjeeling district of West Bengal

Infrastructure	Banekburn Tea Estate (No. of HH = 25)	Dupaka Gaon (No. of HH = 32)	Lamhatta (No. of HH = 20)	Manebhanjgang (No. of HH = 73)	All (No. of HH = 150)
Source of drinking water					
Jhora Water by pump	3 (12)	(0)	(0)	(0)	3 (2)
Rain Water	4 (16)	1 (3.13)	1 (5)	1 (1.37)	7 (4.67)
River water	1 (4)	(0)	(0)	48 (65.75)	49 (32.67)
Spring Water	16 (64)	30 (93.75)	19 (95)	24 (32.88)	89 (59.33)
Tank	1 (4)	1 (3.13)	(0)	(0)	2 (1.33)
Sanitation facility					
No	5 (20)	1 (3.13)	2 (10)	9 (12.33)	17 (11.33)
Yes	20 (80)	31 (96.88)	18 (90)	64 (87.67)	133 (88.67)
Type of Sanitation					
Pucca	5 (25)	9 (29.03)	7 (38.89)	14 (21.88)	35 (23.33)
Normal	13 (65)	22 (70.97)	11 (61.11)	50 (78.13)	96 (64)
Others	2 (10)	0 (0)	0 (0)	0 (0)	2 (1.33)
Accessibility of public Health Care facility					
No	2 (8)	17 (53.13)	13 (65)	3 (4.11)	35 (23.33)
Yes	23 (92)	15 (46.88)	7 (35)	70 (95.89)	115 (76.67)
Banking facility					
No	9 (36)	5 (15.63)	8 (40)	7 (9.59)	29 (19.33)
Yes	16 (64)	27 (84.38)	12 (60)	66 (90.41)	121 (80.67)
Loan facility					
No	20 (80)	17 (53.13)	13 (65)	32 (43.84)	82 (54.67)
Yes through Bank	4 (16)	10 (31.25)	2 (10)	21 (28.77)	37 (24.67)
Yes through Money lender	1 (4)	5 (15.63)	5 (25)	20 (27.4)	31 (20.67)
Housing condition					
Pucca	15 (60)	17 (53.13)	7 (35)	52 (71.23)	91 (60.67)
Semi-pucca	3 (12)	2 (46.25)	(0)	2 (2.74)	7 (4.67)
Kachha	7 (28)	13 (40.63)	13 (65)	19 (26.03)	52 (34.67)

Source Field Survey, 2018; *Note* Figures in the parentheses represent percentage

Table 4.3 Livestock asset per household owned in Darjeeling district

Livestock assets per household	Banekburn Tea Estate (No. of HH = 25)	Dupaka Gaon (No. of HH = 32)	Lamhatta (No. of HH = 20)	Manebhanjgang (No. of HH = 73)	All (No. of HH = 150)
Cow per household	0.47	0.67	0.61	0.12	0.41
Goat per household	0.08	0.13	0.25	–	0.07
Hen per household	10.00	11.10	11.00	6.12	8.51
Pig per household	1.48	0.94	0.75	4.53	2.61

Source Field Survey, 2018

is lowest. Number of cow per household is highest in Dupka Gaon followed by Lamhatta, Banekburn tea estate and Manebhanganj.

Physical assets: The physical asset consists of radio, television (TV) and Mobile for communication; motorbike bi-cycle and car for transportation; refrigerator; LPG connection and stove; Almirah; other furniture comprises chair, table and sofa set (see Table 4.4). 93% and 97% households have TV and mobile phone respectively. It is also observed from Table 4.4 that more than 90% households have LPG connection and cooking stove. This has meaningful implication that the people are not dependent on natural resources like forest for fuel wood. Bicycle is found only in the village

Table 4.4 Distribution of Physical assets in Hill region of Darjeeling District

Physical assets	Banekburn Tea Estate (No. of HH = 25)	Dupaka Gaon (No. of HH = 32)	Lamhatta (No. of HH = 20)	Manebhanjgang (No. of HH = 73)	All (No. of HH = 150)
Radio	4 (16)	9 (28.13)	6 (30)	4 (5.48)	23 (15.33)
T.V	24 (96)	28 (87.5)	17 (85)	71 (97.26)	140 (93.33)
Mobile phone	24 (96)	31 (96.88)	19 (95)	72 (98.63)	146 (97.33)
Motorbike	6 (24)	1 (3.13)	0 (0)	2 (2.74)	9 (6)
Bicycle	(0)	(0)	3 (15)	(0)	3 (2)
Car	3 (12)	1 (3.13)	(0)	5 (6.85)	9 (6)
Refrigerator	13 (52)	14 (43.75)	8 (40)	20 (27.4)	55 (36.67)
LPG connection	24 (96)	31 (96.88)	17 (85)	73 (100)	145 (96.67)
Stove	16 (64)	29 (90.63)	17 (85)	73 (100)	135 (90)
Almirah	24 (96)	32 (100)	20 (100)	73 (100)	149 (99.33)
Other furniture	10 (40)	16 (50)	11 (55)	33 (45.21)	70 (46.67)

Source Field Survey, 2018; *Note* Figures in the parentheses represent percentage

Lamhatta. Walking and public vehicle are the main modes of transportation. 36.67% households have the capability to possess refrigerators.

4.2.1.4 Livelihoods, Income and Consumption of Households in Darjeeling District

Major sources of livelihood of the households in the district of Darjeeling are formal employment (employed in the government sector), driving cum tourist guide, petty business (like tea stall, vending, grocery etc.), casual labour (such as construction labour, labour working to build and maintain roads, drainage, working as porter in tourism sector etc.), Tea Garden labour (who are working in tea garden only) and informal workers (working in hotel, restaurant, shop, shopping mall, factory etc.). About 31% of households are engaged in formal sector like army and banking sector and 28% of households are involved in driving cum tourist guide and 18.67% are engaged in petty business (Table 4.5).

Monthly income per household by livelihood groups across villages in the hill region of Darjeeling district is shown on Table 4.6. Average monthly income per household in the Darjeeling district is Rs. 16282. Among the livelihood groups highest monthly income per household is observed for formal employment (Rs. 25582.85) followed by driver cum tourist guide (Rs. 14339), petty business (Rs. 11644), informal employment (Rs. 10439) and casual labour (Rs. 9929).

Consumption: Table 4.7 depicts that average monthly consumption expenditure per household of the district is Rs. 7558 which is highest in Banekburn Tea estate followed by Lamhatta, Dupka Gaon and Manebhanjganj. Among the livelihood groups monthly consumption expenditure per households is highest for the formally

Table 4.5 Sources of livelihoods in hill region of Darjeeling District

Major sources of livelihoods	Banekburn Tea Estate (No. of HH = 25)	Dupaka Gaon (No. of HH = 32)	Lamhatta (No. of HH = 20)	Manebhanjgang (No. of HH = 73)	All (No. of HH = 150)
Driving/tourist guide	8 (32)	2 (6.25)	6 (30)	26 (35.62)	42 (28)
Petty business	2 (8)	6 (18.75)	5 (25)	15 (20.55)	28 (18.67)
Informal employment	(0)	5 (15.63)	(0)	4 (5.48)	9 (6)
Formal employment	10 (40)	14 (43.75)	6 (30)	17 (23.29)	47 (31.33)
Casual labour	1 (4)	1 (3.13)	2 (10)	10 (13.7)	14 (9.33)
Tea garden labour	4 (16)	4 (12.5)	1 (5)	1 (1.37)	10 (6.67)
Total	25 (100)	32 (100)	20 (100)	73 (100)	150 (100)

Source Field Survey, 2018; *Note* Figures in the parentheses represent percentage

Table 4.6 Monthly income (Rs) per household by livelihood groups in Darjeeling District

Livelihood groups	Banekburn Tea Estate (No. of HH = 25)	Dupaka Gaon (No. of HH = 32)	Lamhatta (No. of HH = 20)	Manebhanjgang (No. of HH = 73)	All (No. of HH = 150)
Driver/tourist guide	18668.70	16353.81	26670.24	10005.71	14338.74
Petty business	14500.48	8727.52	8219.73	13570.67	11643.74
Informal employment	–	9873.29	–	11145.83	10438.86
Formal employment	27517.49	27783.70	21118.67	24207.94	25582.85
Casual labour	11112.58	16865.00	19201.23	7262.17	9928.69
Teagarden labour	6888.28	9906.16	6622.67	4966.67	7876.71
All	19687.65	18121.89	18642.86	13663.22	16282.43

Note Figures in the parentheses represent percentage; *Source* Field Survey, 2018

Table 4.7 Monthly consumption expenditure per household by livelihood groups in Darjeeling district

Livelihood groups	Banekburn Tea Estate (No. of HH = 25)	Dupaka Gaon (No. of HH = 32)	Lamhatta (No. of HH = 20)	Manebhanjgang (No. of HH = 73)	All (No. of HH = 150)
Driver/tourist guide	7808.56	11199.67	12393.75	6628.90	7894.80
Petty business	13514.58	5383.25	7589.40	7677.88	7587.28
work in shop/factory	–	6153.33	–	6929.25	6498.18
Formal employment	10090.00	10452.36	7748.50	7060.12	8803.11
Casual labour	3135.00	3422.00	5288.17	4717.77	4593.65
Teagarden labour	5819.17	5662.50	3438.00	3792.18	5315.68
All	8672.37	8058.45	8640.74	6660.66	7558.15

Note Figures in the parentheses represent percentage; *Source* Field Survey, 2018

employed (Rs. 8803) followed by driver cum tourist guide (Rs. 7895), petty business (Rs. 7587), and informal employee (Rs. 6498) teagarden labour (Rs. 5315) and casual labour (Rs. 4593). The monthly household consumption expenditure for the households living in Manebhanjganj is lowest for all livelihood groups except petty business. Across the villages monthly consumption expenditure per household is highest in Banekburn tea estate which is followed by Lamhatta, Dupka Gaon and Manebhanjganj. In Bankeburn Tea Estate monthly consumption expenditure per household is highest for petty business and lowest for casual labour. In Dupka

Gaon and Lamhatta highest consumption expenditure is found for driver cum tourist guide. In Manebhanjganj monthly consumption expenditure per households is found for formal employees and lowest for Tea Garden Labour.

4.2.2 Description of Sample Households in the Foothill Region of Jalpaiguri District

4.2.2.1 Socio-Economic and Demographic Features of Sample Households of Jalpaiguri District

The primary data was collected from 130 households across three villages in the district Jalpaiguri. Key socio-economic and demographic features of sample households of Jalpaiguri district are shown in Table 4.8.

Social status: Majority households (36.92%) belong to the OBC-A category. Scheduled Tribe (ST) is 25.38% and general category comprises 24.62% of sample households. Mech Basti is tribal dominated village where 91.67% of households belong to ST. In Detha Para three quarter households belongs to OBC-A and 16.22% of households are in general category.

Gender: It shows that 41.54% households are headed by female. It is found from Table 4.8 that 49.12 and 25% households are female headed in the village Gomasta Para and in the village Mech Basti respectively.

Age distribution: Half of households' head are young aged ranging between 21 and 40 years.

Educational status: It is observed that 22.31% households head are illiterate which is highest in the village Mech Basti (33.33%) and lowest in Gomasta Para (12.28%). About 57% households head possess secondary and higher secondary level education. Male are more literate than women in each village of Jalpaiguri district. Female literacy rate is 32.31% while male literacy is 45.38%.

Family size: Average family size is found to be 4.98. Household size is ranging 4.73–5.39.

Economic status: The economic status of the people in foot-hill region of Jalpaiguri is weak. In terms of holding ration card it is found that 67% households have BPL (below poverty line) card which means that they are in the BPL and rest 33% are above poverty line (APL). Highest percentage (89%) of BPL population is found in Mech Basti, followed by Detha Para (70.27%) and Gomasta Para (50.88%).

Landholdings: Land distribution shows that about half households do not possess any cultivable land. Landless households are found be maximum in Gomasta Para (70.18%) and lowest in Mech Basti. 15.38 and 22.31% households belong to marginal farmers and small farmers respectively (land holdings is 1–2 acre).

Table 4.8 Socio-economic conditions of the Households in Foothill region of Jalpaiguri district

Socio-economic variables	Mech Basti (No. of HH = 36)	Detha Para (No. of HH = 37)	Gomasta Para (No. of HH = 57)	All (No. of HH = 130)
Social status				
SC	2 (5.55)	2 (5.41)	13 (22.81)	17 (13.08)
ST	33 (91.67)	–	–	33 (25.38)
General	1 (2.78)	6 (16.22)	25 (43.86)	32 (24.62)
OBC-A	–	29 (78.37)	19 (33.33)	48 (36.92)
Gender				
Male	27 (75.00)	20 (54.05)	29 (50.88)	76 (58.46)
Female	9 (25.00)	17 (45.95)	28 (49.12)	54 (41.54)
Age distribution of head of households				
≤20 years	3 (8.33)	3 (8.10)	3 (5.26)	9 (6.92)
21–40 years	16 (44.44)	18 (48.65)	31 (54.39)	65 (50.00)
41–60 years	15 (41.67)	14 (37.84)	20 (35.09)	49 (37.69)
above 60 years	2 (5.56)	2 (5.41)	3 (5.26)	7 (5.38)
Education				
Illiterate	12 (33.33)	10 (27.03)	7 (12.28)	29 (22.31)
Primary	5 (13.89)	14 (37.84)	7 (12.28)	26 (20.00)
Secondary	14 (38.89)	10 (27.03)	23 (40.35)	47 (36.15)
Above secondary	5 (13.89)	3 (8.10)	20 (35.09)	28 (21.54)
Literacy rate				
Male	18 (50.00)	16 (43.24)	25 (43.86)	59 (45.38)
Female	6 (16.67)	11 (29.73)	25 (43.86)	42 (32.31)
Average of family size	5.39	4.73	4.84	4.98
Economic status				
APL	4 (11.11)	11 (29.73)	28 (49.12)	43 (33.08)
BPL	32 (88.89)	26 (70.27)	29 (50.88)	87 (66.92)
Land holdings(in acre)				
Landless	3 (8.33)	20 (54.05)	40 (70.18)	63 (48.46)
≤1 Acre	6 (16.67)	7 (18.93)	7 (12.28)	20 (15.38)
1 to 2 Acre	15 (41.67)	8 (21.62)	6 (10.53)	29 (22.31)
2 to 3 Acre	8 (22.22)	1 (2.70)	1 (1.75)	10 (7.69)
>3 Acre	4 (11.11)	1 (2.70)	3 (5.26)	8 (6.15)

Note Figures in the parentheses represent percentage, *Source* Field Survey, 2018

4.2.2.2 Infrastructural Facilities of Sample Households in Jalpaiguri District

The infrastructural facilities consist of drinking water facility, sanitation, public health care facility, banking facility, housing etc. of sample households in Jalpaiguri district are displayed in Table 4.9.

Drinking water: "Well" and "Tube Well" are the two prime sources of drinking water in foot-hill region. Most of the households (95%) of Detha Para receives drinking water from Well.

Sanitation facility: The present study finds that 63.08% households have sanitation facility.

Public health care facility: 77% of households have accessibility of public health care facility which is found to be highest in the village of Detha Para (81.08%) and lowest in the village of Mech Basti (75%).

Sources of loan: There are two main sources of loan like bank and money lenders. Only 14% of households have the accessibility of loans from institutional sources like bank. People depend on informal sources of loans like money lenders. It is observed

Table 4.9 Infrastructural facilities in Foothill region of Jalpaiguri district

Infrastructure	Mech Basti (No. of HH = 36)	Detha Para (No. of HH = 37)	Gomasta Para (No. of HH = 57)	All (No. of HH = 130)
Source of drinking water				
Well	30 (83.33)	35 (94.59)	37 (64.91)	102 (78.46)
Tube well	6 (16.67)	2 (5.41)	20 (35.09)	28 (21.54)
Sanitation facility				
Yes	21 (58.33)	16 (43.24)	45 (78.95)	82 (63.08)
No	15 (41.67)	21 (56.76)	12 (21.05)	48 (36.92)
Accessibility of public health care facility				
Yes	27 (75.00)	30 (81.08)	43 (75.44)	100 (76.92)
No	9 (25.00)	7 (18.92)	14 (24.56)	30 (23.08)
Loan facility				
Only from bank	3 (8.33)	4 (10.81)	11 (19.30)	18 (13.85)
Only from money lender	2 (5.56)	14 (37.84)	25 (43.86)	41 (31.54)
No	31 (86.11)	19 (51.35)	21 (36.84)	71 (54.61)
Housing condition				
Pucca	12 (32.43)	37 (64.91)	5 (13.89)	54 (41.54)
Semi-pucca	23 (62.16)	19 (33.33)	1 (2.78)	43 (33.08)
Kachha	2 (5.41)	1 (1.75)	30 (83.33)	33 (25.38)

Note Figures in the parentheses represent percentage, *Source* Field Survey, 2018

that about 31.54% of households take loans from money lenders for meeting various purposes like consumption and production.

Housing condition: Majority of the households (41.54%) of Jalpaiguri district are living with Pucca houses. Semi-pucca house is the second most important dwelling unit where 33.08% households are living and 25% households are living in kachha houses.

4.2.2.3 Households' Assets Structure in Jalpaiguri District

There are two types of assets owned by the households like livestock and physical assets.

Live stock asset: It is found that people in Jalpaiguri district posses mainly hen, pig, goat, and goat etc. as livestock (Table 4.10). Average number of hen per household is highest (1.92) followed by cow (0.81), Goat (0.28), Pig (0.18), buffalo (0.05) and duck (0.02). Average rearing of hen and cow are found highest in Mech Basti and found lowest in Gomasta Para. Rearing of Pig is only found in Mech Basti, a tribal village while rearing of Duck is found only in Gomasta Para.

Physical assets: In Jalpaiguri district physical assets consists of Radio, TV, mobile phone, bi-cycle, refrigerator, LPG, Almirah etc. Distribution of physical assets in foot hill region is given on Table 4.11. It shows that 91.54% and 64.62% of households have mobile phone and TV respectively. Inverter, Refrigerator, sewing machine, Land phone are found only in Gomasta Para whereas stove and van are not found in Gomasta Para. 66.92% of households are possessed bicycle which is the main mode of personnel transport. 56.15% of households owned Almirah. 13.85% of

Table 4.10 Livestock asset per household owned in Jalpaiguri district

Livestock assts per household	Mech Basti (No. of HH = 36)	Detha Para (No. of HH = 37)	Gomasta Para (No. of HH = 57)	All (No. of HH = 130)
Cow per household	0.73	0.60	1.26	0.81
Buffalo per household	0.00	0.02	0.17	0.05
Goat per household	0.16	0.18	0.56	0.28
Hen per household	1.43	1.33	3.36	1.92
Duck per household	–	0.04	0.00	0.02
Pig per household	–	–	0.67	0.18

Source Field Survey, 2018

Table 4.11 Distribution of physical assets in foothill region of Jalpaiguri district

Physical Asset	Mech Basti (No. of HH = 36)	Detha Para (No. of HH = 37)	Gomasta Para (No. of HH = 57)	All (No. of HH = 130)
Radio	1 (2.78)	0 (0.00)	3 (5.26)	4 (3.08)
T.V	21 (58.33)	18 (48.65)	45 (78.95)	84 (64.62)
Mobile phone	27 (75.00)	35 (94.59)	57 (100.00)	119 (91.54)
Land phone	0 (0.00)	0 (0.00)	1 (1.75)	1 (0.77)
Motorbike	9 (25.00)	6 (16.22)	24 (42.11)	39 (30.00)
Bicycle	1 (2.78)	34 (91.89)	52 (91.23)	87 (66.92)
Van	0 (0.00)	1 (2.70)	0 (0.00)	1 (0.77)
Car	1 (2.78)	0 (0.00)	2 (3.51)	3 (2.31)
Cart	1 (2.78)	1 (2.70)	5 (8.77)	7 (5.38)
Tractor	3 (8.33)	1 (2.70)	1 (1.75)	5 (3.85)
Sprayer	1 (2.78)	0 (0.00)	2 (3.51)	3 (2.31)
Pump set	1 (2.78)	1 (2.70)	2 (3.51)	4 (3.08)
Refrigerator	0 (0.00)	0 (0.00)	18 (31.58)	18 (13.85)
LPG connection	10 (27.78)	4 (10.81)	37 (64.91)	51 (39.23)
Sewing machine	0 (0.00)	0 (0.00)	5 (8.77)	5 (3.85)
Stove	3 (8.33)	1 (2.70)	0 (0.00)	4 (3.08)
Inverter	0 (0.00)	0 (0.00)	3 (5.26)	3 (2.31)
Bed	32 (88.89)	37 (100.00)	57 (100.00)	126 (96.92)
Almirah	16 (44.44)	16 (43.24)	41 (71.93)	73 (56.15)
Other furniture	6 (16.67)	21 (56.76)	38 (66.67)	65 (50.00)

Source Field Survey, 2018; *Note* Figures in the parentheses represent percentage

households have the capability to possess refrigerators. 11% of households in the Mech Basti have no Bed and sleeps on the floor. In the three villages only 7 carts are found whereas 5 tractors are available. 39.23% and 3.08% of households have LPG connection and stove respectively. It implies nearly 60% of households are using fuel wood for cooking and heating and dependency on forest is very high.

4.2.2.4 Livelihoods, Income and Consumption of Households in Jalpaiguri District

Major sources of livelihoods in Jalpaiguri district are informal employment,; cultivation, casual labour, employment in the formal sector, and forest product collection. The number of households with different livelihoods is presented on Table 4.12. It shows that 57.69% of households are involved in informal works (such as work in shop, factory, shopping mall etc.), 14.62% of households are engaged in cultivation, 13.08% of households are working as casual labour and 4.62% of households

Table 4.12 Sources of livelihood in foothill region of Jalpaiguri District

Major sources of livelihood	Mech Basti (No. of HH = 36)	Detha Para (No. of HH = 37)	Gomasta Para (No. of HH = 57)	All (No. of HH = 130)
Cultivation	5 (13.51)	7 (12.28)	7 (19.44)	19 (14.62)
Forest product collection	(0)	(0)	6 (16.67)	6 (4.62)
Informal employment	25 (67.57)	33 (57.89)	17 (47.22)	75 (57.69)
Formal employment	1 (2.7)	12 (21.05)	(0)	13 (10)
Casual labour	6 (16.22)	5 (8.77)	6 (16.67)	17 (13.08)
Total	37 (100)	57 (100)	36 (100)	130 (100)

Note Figures in the parentheses represent percentage, *Source* Field Survey, 2018

are collecting non-timber forest product (NTFP). Hence 32% of households have their major occupation as agriculture and allied activities. The rest 10% of households are engaged in formal sector (government employee). Cultivators are found the highest percentage in Mech Basti whereas casual labourers are found to be highest in Dethapara. Forest product collecting households are found only in village of Mech Basti.

Average monthly income per household in the Jalpaiguri district is Rs. 6692 (Table 4.13). Among the livelihood groups highest monthly income per households earns by workers of formal sector followed by cultivators, casual labour, workers of informal sectors and forest product collecting households. Across villages highest monthly household income is found in Gomastapara village followed by village Detha Para and Mech Basti. The households in Gomastapara village earn highest monthly income in all categories of livelihood groups except cultivators. Whereas

Table 4.13 Monthly income per household (Rs) by livelihood groups across villages in Jalpaiguri District

Livelihood groups	Mech Basti (No. of HH = 36)	Detha Para (No. of HH = 37)	Gomasta Para (No. of HH = 57)	All (No. of HH = 130)
Cultivators	954.17	1282.14	1958.93	1445.18
Forest product collectors	–	–	847.22	847.22
Worker in informal sectors	1270.50	1300.76	1073.28	1239.11
Workers in formal sectors	6666.67	59704.86	–	55625.00
Casual labour	1341.94	1636.67	873.61	1263.33
All	1385.18	13623.54	1174.54	6692.90

Note Figures in the parentheses represent percentage, *Source* Field Survey, 2018

Table 4.14 Monthly household's consumption expenditure (Rs) by Livelihood groups across villages in foot hill region of Jalpaiguri District

Livelihood groups	Mech Basti (No. of HH = 36)	Detha Para (No. of HH = 37)	Gomasta Para (No. of HH = 57)	All (No. of HH = 130)
Cultivators	759.40	787.79	787.77	780.31
Forest product collectors	–	–	705.15	705.15
Workers in informal sector	734.01	929.46	787.33	832.09
Workers in formal sector	566.00	953.51	–	923.70
Casual labour	824.45	846.42	759.78	808.09
All	747.57	909.84	769.13	824.69

Note Figures in the parentheses represent percentage, *Source* Field Survey, 2018

households living in Mechbasti are earns lowest monthly household income in all categories of livelihood groups except cultivators.

Consumption: Average monthly consumption expenditure per household of the Jalpaiguri district is Rs. 824.69 which is highest in Gomasta Para followed by Mech Basti and Detha Para (Table 4.14). Among the livelihood group's monthly households consumption expenditure is highest for the workers in formal sector followed by workers in informal sectors, casual labour, cultivators and forest product collecting households. The monthly household consumption expenditure in Gomasta Para is highest for all livelihood groups while monthly household's consumption expenditure living in Detha para is lowest for all livelihood groups (Table 4.14).

4.2.3 Description of Sample Households in the Drought Region of Purulia District

4.2.3.1 Socio-Economic and Demographic Features of Sample Households of Purulia District

The Primary data was collected from 150 households across four villages in the drought prone area of district of Purulia. Key socio-economic and demographic features of sample households are shown in Table 4.15.

Social status: Majority (58%) of households belong to the scheduled tribe (ST) who are most disadvantageous section of the society. ST consists highest percentage in the village Matha (69.70%) followed by village Ajodhya (63.64%), Banduri (50.98%) and Ebildi (45.45%).

Table 4.15 Socio-economic conditions of the households in Purulia district

Socio-economic variables	Ajodhya (No. of HH = 44)	Ebildi (No. of HH = 22)	Matha (No. of HH = 33)	Banduri (No. of HH = 51)	All (No. of HH = 150)
Social status					
SC	10 (22.72)	2 (9.09)	2 (6.06)	14 (27.45)	28 (18.67)
ST	28 (63.64)	10 (45.45)	23 (69.70)	26 (50.98)	87 (58.00)
General	4 (9.09)	1 (4.55)	8 (24.24)	5 (9.80)	18 (12.00)
OBC	2 (4.55)	9 (40.91)	0 (0.00)	6 (11.76)	17 (11.33)
Gender					
Male	32 (72.73)	20 (90.91)	26 (78.79)	39 (76.47)	117 (78.00)
Female	12 (27.27)	2 (9.09)	7 (21.21)	12 (23.53)	33 (22.00)
Age distribution of head of households					
≤20 years	1 (2.27)	1 (4.56)	0 (0.00)	3 (5.88)	5 (3.33)
21–40 years	21 (47.73)	7 (31.82)	15 (45.45)	27 (52.94)	70 (46.67)
41–60 years	17 (38.64)	9 (40.91)	16 (48.48)	13 (25.49)	55 (36.67)
above 60 years	5 (11.36)	5 (22.73)	2 (6.06)	8 (15.69)	20 (13.33)
Education					
Illiterate	18 (40.91)	12 (54.55)	7 (21.21)	24 (47.06)	61 (40.67)
Primary	3 (6.82)	5 (22.73)	7 (21.21)	5 (9.80)	20 (13.33)
Secondary	17 (38.64)	5 (22.73)	13 (39.39)	18 (35.29)	53 (35.33)
Above secondary	6 (13.64)	0 (0.00)	6 (18.18)	4 (7.84)	16 (10.67)
Literacy rate					
Male	21 (47.73)	8 (36.36)	15 (45.45)	13 (25.49)	57 (38.00)
Female	5 (11.36)	2 (9.09)	1 (3.03)	5 (9.80)	13 (8.67)
Family size	4.64	5.55	4.18	5.02	4.80
Economic status					
APL	0 (0.00)	1 (4.55)	1 (3.03)	0 (0.00)	2 (1.33)
BPL	44 (100.00)	21 (95.45)	32 (96.97)	51 (100.00)	148 (98.67)
Land holding (acre)					
Landless	8 (18.18)	4 (18.18)	11 (33.33)	6 (11.76)	29 (19.33)
≤1 acre	22 (50.00)	13 (59.09)	7 (21.21)	39 (76.47)	81 (54.00)
1.00001 to 2 acre	8 (18.18)	2 (9.09)	10 (30.30)	4 (7.84)	24 (16.00)
2.00001 to 3 acre	4 (9.09)	3 (13.64)	2 (6.06)	1 (1.96)	10 (6.67)
>3 acre	2 (4.55)	0 (0.00)	3 (9.09)	1 (1.96)	6 (4.00)

Source Field Survey, 2018; *Note* Figures in the parentheses represent percentage

Gender: 22% of households are headed by female. Concentration of female headed households in village Ebilidi is very low as 9% whereas maximum in the village Ajodhya (27.27%).

Age distribution: Majority (46.67%) of households' head are ranging between 21 to 40 years. This age distribution follows all the villages except Matha where majority (48.48%) of head of the households belongs to the middle age category 41–60 years.

Educational status: It shows that 40.67% head of the households are illiterate. Illiteracy is highest in village Ebildi (54.55%) followed by Banduri, Ajodhya and Matha. About 36.67% households head possess secondary level education. Another 10.67% of head of the households are educated above secondary. Again it is observed that female literacy in Purulia is very low as 8.67%. Female literacy rate is maximum (11.36%) in Ajodhya village. The educational status shows that basic education of Purulia district is very weak. It indicates poor status of human capital in the district of Purulia.

Family size: Average number of family member in each household in Purulia district is 4.80. Household size is observed highest in Ebildi (5.55) and lowest in Matha (4.18).

Economic status: In terms of BPL card holding, almost all households (98.67%) are lying below poverty line (BPL) in drought prone area of Purulia district. In Banduri and Ajodhya village, all households are BPL. Land holding status shows that one-fifth households possess no cultivable land. Across the villages' maximum land less households are found in Matha (33.33%). In Purulia district majority (54%) households have less than one acre cultivable land which is highest in Banduri (76.47%) and lowest in Matha (21.21%). This implies that 75% of sample households are constituted by land less and marginal farmers.

4.2.3.2 Infrastructural Facilities of Sample Households in Drought Region of Purulia District

Analysis of infrastructural facilities of sample households in Purulia district is presented in Table 4.16.

Drinking water: It shows that "Tube well" and "Well" are the main source of drinking water in drought prone Purulia district. Dependency for drinking water from "Well" is very low in Ajodhya (4.55%) and highest in Banduri (37.25%).

Sanitation facility: Majority of the households (78.67%) have no access to sanitation facility. Absent of sanitation facility is highest both in the villages of Ajodhya and Banduri (90.91%) and lowest in the village of Ebildi (45.45%).

Public health care facility: Most of the households (90%) have accessibility to public health care facility which is highest in the village of Ajodhya (93.18%) and lowest in the village of Matha (78.79%).

Table 4.16 Infrastructural facilities in Drought Prone Purulia District of West Bengal

Infrastructure	Ajodhya (No. of HH = 44)	Ebildi (No. of HH = 22)	Matha (No. of HH = 33)	Banduri (No. of HH = 51)	All (No. of HH = 150)
Source of drinking water					
Well	2 (4.55)	17 (77.27)	10 (30.30)	19 (37.25)	48 (32.00)
Tube well	42 (95.45)	5 (22.73)	23 (69.70)	32 (62.75)	102 (68.00)
Sanitation facility					
Yes	4 (9.09)	12 (54.55)	3 (9.09)	13 (25.49)	32 (21.33)
No	40 (90.91)	10 (45.45)	30 (90.91)	38 (74.51)	118 (78.67)
Accessibility of public Health Care facility					
Yes	41 (93.18)	19 (86.36)	26 (78.79)	49 (96.08)	135 (90.00)
No	3 (6.82)	3 (13.67)	7 (21.21)	2 (3.92)	15 (10.00)
Loan facility					
No	33 (75)	47 (92.16)	10 (45.45)	24 (72.73)	114 (76)
yes	11 (25)	4 (7.84)	12 (54.55)	9 (27.27)	36 (24)
Not from any source	44 (100.00)	22 (100.00)	33 (100.00)	51 (100.00)	150 (100.00)
Housing condition					
Pucca	3(6.82)	3(5.88)	0	2(6.06)	8(5.33)
Semi-pucca	41(93.18)	48(94.12)	22(100)	30(90.91)	141(94)
Kachha	0	0	0	1(3.03)	1(0.67)

Source Field Survey, 2018; *Note* Figures in the parentheses represent percentage

Sources of loan: The study observes that 76% of sample households do not have institutional/non institutional credit facility. In Matha village maximum households (54.55%) got institutional loan while it was 7.68% in Ebildi village.

Housing condition: It is observed that most of the households (94%) are living in semi-pucca houses in drought prone Purulia district. In Matha village, no pucca house was found.

4.2.3.3 Households' Assets Distribution in the District of Purulia

There are two types of assets viz. livestock and physical assets are taken into the analysis.

Live stock asset: It is found that people in Purulia district posses mainly hen, goat, cow, buffalo duck and pig etc. as livestock (Table 4.17). Average number of hen per households owned is highest (2.66) followed by Goat (1.51), Cow (0.83), Duck (0.09), Buffalo (0.07) and Pig (0.05). Average rearing of Goat is highest in the village of Matha (2.55) and lowest in the village of Banduri (1.02) village. Average rearing of hen is highest in the village of Ajodhya (2.98) and lowest in Banduri (2.00) village.

Table 4.17 Village wise average livestock asset per household owned in Drought Prone Purulia District

Livestock assets per household	Ajodhya (No. of HH = 44)	Ebildi (No. of HH = 22)	Matha (No. of HH = 33)	Banduri (No. of HH = 51)	All (No. of HH = 150)
Cow per household	0.45	1.10	1.50	0.48	0.83
Buffalo per household	0.05	0.08	0.18	0.00	0.07
Goat per household	1.48	1.02	1.18	2.55	1.51
Hen per household	2.98	2.00	3.14	2.94	2.66
Duck per household	0.02	0.10	0.00	0.24	0.09
Pig per household	0.07	0.00	0.09	0.06	0.05

Source Field Survey, 2018

No household in Banduri village possess pig. Households of village Ebildi possess no duck whereas households of Matha village posses no pig.

Physical assets: Physical assets consist of television, mobile phone, LPG connection, cooking stove and almirah etc. (Table 4.18). It is revealed that 82% of households have mobile phone. Bicycle is the prime mode of private transportation. 59.33% households in Purulia posses bicycle. Possession of motorbike in Purulia district is limited (8.67%). Only 10% household's posses cart. Pump set for irrigation is owned by only 6% households. There are no households with LPG connection in Ebldi and Banduri village. LPG connection in other two villages have is very low i.e. 2.27% of households in Ajodhya and 6.06% of households in Matha. It indicates that people are highly dependent on natural resources like fuel wood for cooking purposes.

4.2.3.4 Livelihoods, Income and Consumption of Households in Purulia District

The major sources of livelihoods of the households in Purulia district are agriculture; informal employment; forest product collection and casual work (see Table 4.19). One third of households are engaged in cultivation. Cultivators are found to be highest in the villages of Ajodhya and Matha. Second important livelihoods group is workers involved in informal sector (28%) which is found be highest in the village of Ebildi. Third important livelihoods group is forest dependent people (26%) who are found to be highest in Banduri and Matha villages. About 13.33% of households are casual labour.

Table 4.18 Distribution of physical assets in drought prone Purulia district

Physical assets	Ajodhya (No. of HH = 44)	Ebildi (No. of HH = 22)	Matha (No. of HH = 33)	Banduri (No. of HH = 51)	All (No. of HH = 150)
T.V	2 (4.55)	0 (0.00)	7 (21.21)	4 (7.84)	13 (8.67)
Mobile phone	34 (77.27)	19 (86.36)	27 (81.82)	43 (84.31)	123 (82.00)
Motorbike	5 (11.36)	2 (9.09)	2 (6.06)	4 (7.84)	13 (8.67)
Bicycle	22 (50.00)	13 (59.09)	19 (57.58)	35 (68.63)	89 (59.33)
Cart	4 (9.09)	4 (18.18)	2 (6.06)	5 (9.80)	15 (10.00)
Tractor	0 (0.00)	1 (4.55)	0 (0.00)	0 (0.00)	1 (0.67)
Cultivator	1 (2.27)	0 (0.00)	0 (0.00)	6 (11.76)	7 (4.67)
Sprayer	2 (4.55)	0 (0.00)	2 (6.06)	2 (3.92)	6 (4.00)
Threshing machine	2 (4.55)	1 (4.55)	3 (9.09)	2 (3.92)	8 (5.33)
Pump set	3 (6.82)	2 (9.09)	1 (3.03)	3 (5.88)	9 (6.00)
LPG connection	1 (2.27)	0 (0.00)	4 (12.12)	0 (0.00)	5 (3.33)
Stove	1 (2.27)	0 (0.00)	2 (6.06)	0 (0.00)	3 (2.00)
Bed	44 (100.00)	17 (77.27)	30 (90.91)	48 (94.12)	139 (92.67)
Almirah	5 (11.36)	2 (9.09)	9 (27.27)	1 (1.96)	17 (11.33)
Other furniture	2 (4.55)	3 (13.64)	3 (9.09)	0 (0.00)	8 (5.33)

Source Field Survey, 2018; *Note* Figures in the parentheses represent percentage

Table 4.19 Sources of livelihoods in drought pone Purulia district

Major sources of livelihoods	Ajodhya (No. of HH = 44)	Ebildi (No. of HH = 22)	Matha (No. of HH = 33)	Banduri (No. of HH = 51)	All (No. of HH = 150)
Agriculture	18 (40.91)	12 (23.53)	6 (27.27)	13 (39.39)	49 (32.67)
Forest product collection	11 (25)	14 (27.45)	5 (22.73)	9 (27.27)	39 (26)
Informal employment	13 (29.55)	14 (27.45)	7 (31.82)	8 (24.24)	42 (28)
Casual labour	2 (4.55)	11 (21.57)	4 (18.18)	3 (9.09)	20 (13.33)
	44 (100)	51 (100)	22 (100)	33 (100)	150 (100)

Source Field Survey, 2018; *Note* Figures in the parentheses represent percentage

Monthly income per household of different livelihood groups in Purulia district is displayed in Table 4.20. It observed from this table that average monthly income per household in Purulia district is Rs. 809.21. Households living in Ebildi village are earning the highest income (Rs. 850.83) while households living in Matha village are earning the lowest income (Rs. 761.87). Among the livelihood groups highest

Table 4.20 Monthly income (Rs) per households of different livelihood groups in drought prone Purulia District

Livelihood groups	Ajodhya (No. of HH = 44)	Ebildi (No. of HH = 22)	Matha (No. of HH = 33)	Banduri (No. of HH = 51)	All (No. of HH = 150)
Cultivators	864.12	877.01	907.92	814.10	859.37
Forest dependent people	752.88	832.74	806.67	723.15	781.58
Informal worker	703.21	754.36	892.26	700.00	751.15
Casual labour	837.50	920.45	747.92	816.67	862.08
	787.56	840.56	850.83	761.87	809.21

Source Field Survey, 2018; *Note* Figures in the parentheses represent percentage

income is earned by casual labour (Rs. 862.08) followed by cultivators (Rs. 859.37), forest dependent people (Rs. 781.58) and workers in informal sector (Rs. 751.08).

Monthly consumption expenditure per household of livelihood groups in drought region of Purulia district is displayed in Table 4.21. It is observed that average monthly consumption expenditure per household in Purulia district is found to be Rs726.68. Households living in Banduri villages are spending the highest (Rs. 780.08) for consumption while households living in Matha villages are spending the lowest (Rs. 643.04) for consumption. Among the livelihood group's highest expenditure for consumption is borne by households with occupation of casual labour (Rs. 862) followed by cultivators (Rs. 859), forest dependent people (Rs. 781.58), workers in the informal sector (Rs. 751).

Table 4.21 Monthly consumption expenditure (Rs) household of different livelihood groups in drought prone Purulia district

Livelihood groups	Ajodhya (No. of HH = 44)	Ebildi (No. of HH = 22)	Matha (No. of HH = 33)	Banduri (No. of HH = 51)	All (No. of HH = 150)
Cultivators	719.72	826.93	759.67	628.00	726.53
Forest dependent people	728.26	790.52	752.02	681.93	742.97
Workers in informal sectors	674.85	733.14	784.36	566.13	691.82
Casual labour	795.10	775.43	715.03	796.67	768.50
	712.03	780.08	757.67	643.04	726.68

Source Field Survey, 2018; *Note* Figures in the parentheses represent percentage

4.2.4 Description of Sample Households in the Coastal Sundarban in the District of South 24 Parganas

4.2.4.1 Socio-Economic and Demographic Features of Sample Households at Costal Sundarban of South 24 Parganas District

Social status: Majority of households (52.29%) belong to the Scheduled Caste (SC). Scheduled Tribe (ST) population is the second major social group consisting 27.42% households. ST population has concentrated mainly in the village of Paschim Dwarikapur. Other backward caste (OBC) composed of one fifth of the sample households.

Gender: It is observed that one third of sample households are headed by female. The concentration of female headed households is found to be in the village of Madhabnagar where half of sample households are female headed. In Bhagbatpur and Paschim Dwarikapur villages female headed households are lower than the other villages.

Age distribution: Average age of head of sample households is 43.66 years. More than 86% head of households belong to working age group like 21–60 years.

Educational status: Educational status shows that more than half of sample household's heads are illiterate. Illiteracy is found to be the highest in Paschim Dwarikapur where 84.13% of household heads are illiterate. Second most illiteracy is found in the village of Madhab Nagar (64.58%). About 30% of household's head possess secondary level education. Another 13.71% of household heads are educated up to primary level.

Family size: Average number of members in each household of coastal sundarban is 3.88.

Economic status: As per ration card issued by the government of West Bengal, majority households (74.11%) are laying below poverty line (BPL) and rest 25.89 belong to above poverty line. In Paschim Dwarikapur village, all households belong to BPL category (Table 4.22).

4.2.4.2 Infrastructural Facilities of Sample Households in of Costal Sundarban, South 24 Parganas District

Analysis of infrastructural facilities of sample households at costal Sundarban in South 24 Parganas district is displayed in Table 4.23.

Drinking water: It shows that "Tube well" is the prime source of drinking water. 60% of sample households are using tube well for safe drinking water. The rest 40% people have to drink river water during crab collection or fishing.

Table 4.22 Socio-economic conditions of the households at costal Sundarban of S24 Parganas district

Socio-economic variables	Bhagbatpur (No. of HH = 30)	Laxmi Narayanpur (No. of HH = 68)	Madhabnagar (No. of HH = 48)	Paschim Dwarikapur (No. of HH = 51)	All (No. of HH = 197)
Social status					
SC	28 (93.34)	30 (44.12)	44 (91.67)	1 (1.97)	103 (52.29)
ST	1 (3.34)	1 (1.48)	2 (4.17)	50 (98.04)	54 (27.42)
General	1 (3.34)	0	0	0	1 (0.51)
OBC	0	37 (54.42)	2 (4.17)	0	39 (19.8)
Gender					
Female	4 (13.34)	28 (41.18)	24 (50)	9 (17.65)	65 (33)
Male	26 (86.67)	40 (58.83)	24 (50)	42 (82.36)	132 (67.01)
Education					
Illiterate	10 (33.34)	16 (23.53)	31 (64.59)	43 (84.32)	100 (50.77)
Primary	2 (6.67)	17 (25)	7 (14.59)	1 (1.97)	27 (13.71)
Secondary	16 (53.34)	28 (41.18)	9 (18.75)	6 (11.77)	59 (29.95)
Above secondary	2 (6.67)	7 (10.3)	1 (2.09)	1 (1.97)	11 (5.59)
Age distribution of head of households					
≤20 years	1 (3.23)	0	3 (6.25)	0	4 (2.04)
21–40 years	12 (38.71)	33 (49.26)	19 (39.59)	21 (41.18)	85 (43.15)
41–60 years	14 (45.17)	28 (41.8)	16 (33.34)	27 (52.95)	85 (43.15)
>60 years	4 (12.91)	6 (8.96)	10 (20.84)	3 (5.89)	23 (11.68)
Average of age of head of households (Years)	46.53	41.28	45.21	43.71	43.66
Illiteracy rate					
Female	1 (3.34)	7 (10.3)	19 (39.59)	8 (15.69)	35 (17.77)
Male	9 (30)	9 (13.24)	12 (25)	35 (68.63)	65 (33)

(continued)

Table 4.22 (continued)

Socio-economic variables	Bhagbatpur (No. of HH = 30)	Laxmi Narayanpur (No. of HH = 68)	Madhabnagar (No. of HH = 48)	Paschim Dwarikapur (No. of HH = 51)	All (No. of HH = 197)
Economic status					
BPL	19 (63.34)	57 (83.83)	19 (39.59)	51 (100)	146 (74.12)
APL	11 (36.67)	11 (16.18)	29 (60.42)	0	51 (25.89)
Land holding (Acre)					
Landless	10 (33.34)	36 (52.95)	36 (75)	51 (100)	133 (67.52)
≤1 Acre	12 (40)	27 (39.71)	10 (20.84)	0 (0)	49 (24.88)
1.00001–2 Acre	8 (26.67)	4 (5.89)	2 (4.17)	0 (0)	14 (7.11)
2.00001–3 Acre	0	1 (1.48)	0	0	1 (0.51)
Average of family size	4.03	4.01	3.42	4.04	3.88

Source Field Survey, 2018; *Note* Figures in the parentheses represent percentage

Source of light: Majority (95%) of households has electricity connection and the rest of the households are using kerosene lamp for lighting their houses.

Sanitation facility: The present study finds that more than 53% households have no access to the sanitation facility. It indicates poor status of development. Absence of sanitation facility is highest in Laxmi Narayanpur villages where more than three fourth villagers do not have any toilet.

Public health care facility: Majority (52.8%) of sample households in Sundarban having inaccessibility of public health care facility. Maximum (69.12%) inaccessibility of primary health care facility is found in Laxmi Narayanpur village.

Banking facility: The study observes that on average 58% sample households have banking facility. No villagers in Paschim Dwarikapur have financial inclusion with the bank. Villagers of Laxmi Narayanpur (97.05%) enjoy maximum banking facility followed by Bhagbatpur (86.66%) and Madhabnagar (45.83%).

Mode of transport: Most important mode of transport in costal Sundarban is bicycle followed by Vanrikshaw, Boat and scooter. 42.63% households are moving here and there with the help of personnel bi-cycle. The highest percent of bi-cycle user is found in Madhabnagar (97.91%) and lowest in Paschim Dwarikapur (2%). No persons of Paschim Dwarikapur are using van-rickshaw and scooter while 98% people of this village are using boat for their movement during crab collection. The maximum van rickshaw users are observed in Laxmi Narayanpur village (73.52%).

Table 4.23 Infrastructural facilities in costal Sundarban, S24 Parganas District of West Bengal

Infrastructure	Bhagbatpur (No. of HH = 30)	Laxmi Narayanpur (No. of HH = 68)	Madhabnagar (No. of HH = 48)	Paschim Dwarikapur (No. of HH = 51)	All (No. of HH = 197)
Source of light in house					
Electricity	29 (96.67)	68 (100)	48 (100)	43 (84.32)	188 (95.44)
Kerosene	1 (3.34)	0	0	8 (15.69)	8 (4.07)
Sources of drinking water					
River Water	3 (10)	19 (27.95)	5 (10.42)	51 (100)	78 (39.6)
Tube well	27 (90)	49 (72.06)	43 (89.59)	0	119 (60.41)
Sanitation facility					
NO	5 (16.67)	52 (76.48)	24 (50)	24 (47.06)	105 (53.3)
Yes	25 (83.34)	16 (23.53)	24 (50)	27 (52.95)	92 (46.71)
Accessibility of public health centre					
NO	4 (13.34)	47 (69.12)	27 (56.25)	26 (50.99)	104 (52.8)
Yes	26 (86.67)	21 (30.89)	21 (43.75)	25 (49.02)	93 (47.21)
Availability of Banking facility in locality					
NO	4 (13.34)	2 (2.95)	26 (54.17)	51 (100)	83 (42.14)
Yes	26 (86.67)	66 (97.06)	22 (45.84)	0	114 (57.87)
Mode of transport					
Bi-Cycle	19 (63.34)	17 (25)	47 (97.92)	1 (1.97)	84 (42.64)
Scooter	2 (6.67)	1 (1.48)	1 (2.09)	0	3 (1.53)
Van-rickshaw	9 (30)	50 (73.53)	0	0	59 (29.95)
Boat	0	0	0	50 (98.04)	50 (25.39)
Housing condition					
Pucca	20 (66.67)	20 (29.42)	2 (4.17)	0	42 (21.32)
Semi Pucca	0	23 (33.83)	0	10 (19.61)	33 (16.76)
Kachha	10 (33.34)	25 (36.77)	46 (95.84)	41 (80.4)	122 (61.93)

Source Field Survey, 2018; *Note* Figures in the parentheses represent percentage

Housing condition. Majority of sample households in the present study are living in Katcha houses (62%), which is followed by Pucca houses (21%) and Semi Pucca houses (17%). In Paschima Dwarikapur no houses are pucca. Katcha houses are found to be highest in the village of Madhab Nagar (95.84%), followed by Paschima Dwarikapur (80.4%), Laxmi Narayanpur (36.77%) and Bhagbatpur (33.34%).

4.2.4.3 Households Assets of Costal Sundarban of South 24 Parganas District

Two types of assets viz. livestock assets and physical assets are taken into the analysis.

Live stock asset: Livestock assets per households in coastal Sundarban are presented in Table 4.24. It is found from the Table 4.24 that people in coastal Sundarban posses mainly hen, duck, cow, goat and pig etc. as livestock. Average number of duck per household owned is highest (4.50) followed by Hen (4.15), Pig (0.46), Cow (0.42) and Goat (0.22). Rearing of Duck is found highest in village Laxmi Narayanpur followed by Bhagbatpur and Madhabnagar. While rearing of Hen is found highest in Bhagbatpur followed by Laxmi Narayanpur and Madhabnagar. Rearing of Pig is found only in Paschim Dwarikapur where each household possesses 1.78 pigs. Cow is available maximum in village Madhabnagar followed by Bhagbatpur and Laxmi Narayanpur. Rearing of Goat is found only in Laxmi Narayanpur and Bhagbatpur.

Physical: Physical asset of the households in coastal Sundarban consists of TV, mobile phone, van, bi-cycle, LPG connection, stove, fishing net, boat, almirah and other furniture comprises chair, table and sofa set. About 82.75% of households have mobile phone. The second and third important assets possessed by the people are bi-cycle and TV respectively. Possession of fishing net and boat is found to be maximum in Laxminarayanpur. Motorbike, van, refrigerator, pumpset, LPG connection almirah are not found in Paschim Dwarikapur. Pumset is found only in Laxmi Narayanpur village (Table 4.25).

Table 4.24 Village wise average number of livestock asset per household owned in Costal Sundarban, S24 Parganas District

Livestock assets per household	Bhagbatpur (No. of HH = 30)	Laxmi Narayanpur (No. of HH = 68)	Madhabnagar (No. of HH = 48)	Paschim Dwarikapur (No. of HH = 51)	All (No. of HH = 197)
Cow per household	0.40	0.37	0.88	0.06	0.42
Goat per household	0.30	0.51	0.00	0.00	0.22
Hen per household	6.27	5.53	5.27	0.00	4.15
Duck per household	6.50	7.91	3.21	0.00	4.50
Pig per household	0.00	0.00	0.00	1.78	0.46

Source Field Survey, 2018

Table 4.25 Village wise distribution of Physical in Costal Sundarban, South 24 Parganas district

Physical	Bhagbatpur (No. of HH = 30)	Laxmi Narayanpur (No. of HH = 68)	Madhabnagar (No. of HH = 48)	Paschim Dwarikapur (No. of HH = 51)	All (No. of HH = 197)
Radio	2 (6.67)	7 (10.3)	8 (16.67)	2 (3.93)	19 (9.65)
T.V	29 (96.67)	14 (20.59)	19 (39.59)	18 (35.3)	80 (40.61)
Mobile phone	24 (80)	55 (80.89)	43 (89.59)	41 (80.4)	163 (82.75)
Bicycle	29 (96.67)	31 (45.59)	34 (70.84)	13 (25.5)	107 (54.32)
Motor bike	2 (6.67)	1 (1.48)	3 (6.25)	0 (0)	6 (3.05)
Van	1 (3.34)	2 (2.95)	4 (8.34)	0	7 (3.56)
Refrigerator	1 (3.34)	1(1.48)	0	0	2 (1.02)
Pump set	0	1 (1.48)	0	0	1 (0.51)
LPG connection	6 (20)	11 (16.18)	10 (20.84)	0	27 (13.71)
Stove	0	5 (7.36)	0	6 (11.77)	11 (5.59)
Bed	12 (40)	9 (13.24)	1 (2.09)	17 (33.34)	39 (19.8)
Almirah	9 (30)	2 (2.95)	1 (2.09)	0	12 (6.1)
Fishing net	2 (6.67)	34 (50)	4 (8.34)	2 (3.93)	42 (21.32)
Boat	0	28 (41.18)	4 (8.34)	10 (19.61)	42 (21.32)
Other	0	1 (1.48)	1 (2.09)	16 (31.38)	18 (9.14)
All	30 (100)	68 (100)	48 (100)	51 (100)	197 (100)

Source Field Survey, 2018; *Note* Figures in the parentheses represent percentage

4.2.4.4 Livelihood, Income and Consumption of Households of Costal Sundarban in South 24 Parganas District

Major sources of livelihoods of the households in the coastal Sundarban are fishing, crab collection, petty business and casual labour (Table 4.26). Most important livelihoods group is workers in informal sector (worker in rice mill or cold storage) because maximum (31%) households of coastal Sundarban are engaged in this occupation. Second most important livelihoods group is crab collection. 26.4% households are engaged in this category. Third most important livelihoods group is casual labour in which 19.29% households are engaged. Fourth most livelihoods group is fishing where 14.22 sample households are drawing their livelihoods. The least important (9.14%) major source of livelihood is petty business such as hawker, small shop, fish or crab arat etc. In Bhagbatpur majority of people are doing petty business of fish arat. In Laxminarayanpur village one third of households are engaged in each livelihoods such as informal employment, fishing and casual labourer. In Madhabnagar two third of populations are working in shop or factory outside the locality and one

Table 4.26 Sources of livelihoods at coastal Sundarban of South 24 Parganas district

Major sources of livelihoods	Bhagbatpur N = 30	Laxmi Narayanpur N = 68	Madhabnagar N = 48	Paschim Dwarikapur N = 51	All N = 197
Fishing	3 (10)	20 (29.42)	5 (10.42)	–	28 (14.22)
Informal employment	8 (26.67)	23 (33.83)	30 (62.5)	–	61(30.97)
Petty business	11 (36.67)	6 (8.83)	1 (2.09)	–	18 (9.14)
Crab collection	1 (3.34)	–	–	51 (100)	52 (26.4)
Casual labour	7 (23.34)	19 (27.95)	12 (25)	–	38 (19.29)

Source Field Survey, 2018; *Note* Figures in the parentheses represent percentage

fourth households are doing casual labourer. In Paschim Dwarikapur village's sole occupation is crab collection.

Average monthly income per household in coastal Sundarban is Rs. 6113.79 which is found maximum in Bhagbatpur village (Rs. 7321.33) and followed by village Laxmi Narayanpur (Rs. 6459.47), Paschim Dwarikapur (Rs. 5548.90) and Madhabnagar (Rs. 5469.56) (Table 4.27). Monthly income per household is observed to be highest (Rs. 8763.39) for the livelihood group of petty business men. The lowest monthly income per household is derived by the casual labourer (Rs4790.29) (Table 4.27).

Consumption: Monthly consumption expenditure per household by livelihood groups across villages at costal Sundarban, S 24 Parganas district is displayed in

Table 4.27 Average monthly income of the households by livelihood groups in the different villages in S 24 Parganas

Livelihood groups	Bhagbatpur	Laxmi Narayanpur	Madhabnagar	Paschim Dwarikapur	All
Fishing communities	5296.67	6219.95	5000	–	5903.18
Workers in the informal sector	6884.88	7638.3	5984.23	–	6726.02
Petty business men	10031.3	7381	3111	–	8763.39
Crab collecting communities	6075	–	–	5548.9	5559.02
Casual labour	4607.43	4993.58	–	–	4790.29
All	7321.33	6459.47	5469.56	5548.9	6113.79

Source Field Survey, 2018

Table 4.28 Monthly consumption expenditure (Rs) per household of livelihood groups at coastal Sundarban of south 24 Parganas

Livelihood groups	Bhagbatpur	Laxmi Narayanpur	Madhabnagar	Paschim Dwarikapur	All
Fishing communities	3545.67	2503.9	2610.2	–	2634.5
Workers in the informal sector	3251.88	2369.09	2557.4	–	2577.48
Petty business men	2549.36	2102.83	2503	–	2397.94
Crab collecting communities	2451	–	–	1596.45	1612.88
Casual labourer	2664.43	2652.68	2700.17	–	2669.84
All	2859.9	2464.49	2597.46	1596.45	2332.38

Source Field Survey, 2018; *Note* Figures in the parentheses represent percentage

Table 4.28. It is observed that average monthly consumption expenditure per households in Sundaban is Rs. 2332.38. Of the village the monthly consumption expenditure per household is found to be highest in the village Bhagbatpur and lowest for Paschim Dwarikapur. Of the livelihood groups the monthly consumption expenditure per household is found to be highest for casual labourer while it is lowest for crab collecting households (Table 4.28).

4.2.5 Description of Sample Households in the Coastal Region of East Midnapore District

In this section we describe the socio-economic-demographic features, infrastructural facilities, assets structure, livelihoods, income and consumption of sample households in the coastal region of East Midnapore district.

4.2.5.1 Description of Sample Households in Coastal Region of East Midnapore District

Social status: Majority (93%) of the households belong to the general caste and rest 7% of households belong to the scheduled caste (SC). Percentages of general caste household are higher in the village of Purba Mukundapur while SC households are more in the village of Maitrapur.

Gender: 85% of households are male headed while 15% of households are female headed in the district of East Midnapore.

Age: Average age of the household's head in East Midnapore district is 45 years. Majority (44.03%) of the heads' of household are middle aged ranging between 41 and 60 years.

Educational status: In the East Midnapore district about 92% of households are literate. About 50% of household's head possess secondary level of education and another 22% household's head are above higher secondary level of education.

Household size: Average number of family size is 3.89.

Economic status: It is observed that majority (59%) of the households are lying above poverty line (APL) and the rest 41% of households are lying below poverty line.

Land holdings: It is observed that that 51% of households of coastal East Midnapore district are landless. About 44.65% of households possess less than 1 acre of cultivable land. Landless households are found to be more in Maitrapur whereas marginal land holding households are living in the village of Purba Mukundapur (Table 4.29).

4.2.5.2 Infrastructural Facilities of Sample Households in the Coastal Region of East Midnapore District

The infrastructural facilities like drinking water, sanitation, health care and banking etc. of sample households in coastal region of East Midnapore district are presented in Table 4.30.

Drinking water: It shows that Tube well is the prime source of drinking water followed by Pump well, river water and pipe water. About 93% of households use tube well as their source of drinking water.

Sanitation facility: 78.62% of households have access to the sanitation facility in East Midnapore.

Public health care facility: 85% of households have access to public health care facility.

Banking facility: More than 75% of households have access to banking facility.

Housing condition: It is observed that about 38.36% of households reside in the pucca houses. About 33.96% of households are living in semi-pucca houses and rest 27.67% of households are living in kuchha houses. Only 19%% of households are the beneficiaries of Indira Ayas Yojana (IAY), government housing scheme for the poor.

Table 4.29 Socio-Economic conditions of the households in coastal region of East Midnapore district

Socio-economic variables	Maitrapur (No. of HH = 89)	Purba Mukundapur (No. of HH = 70)	All (No. of HH = 159)
Social status			
SC	7 (7.87)	4 (5.71)	11 (6.92)
Gen	82 (92.13)	66 (94.29)	148 (93.08)
Gender			
female	16 (17.98)	8 (11.43)	24 (15.09)
Male	73 (82.02)	62 (88.57)	135 (84.91)
Age distribution of head of households			
Average age of household head (Years)	46.03	44.80	45.49
21–40 years	36(40.45)	30(42.86)	66(41.51)
41–60 years	39 (43.82)	31 (44.29)	70 (44.03)
above 60 years	14 (15.73)	9 (12.86)	23 (14.47)
Education			
Illiterate	5 (5.62)	7 (10)	12 (7.55)
Primary	21 (23.6)	11 (15.71)	32 (20.13)
secondary	40 (44.94)	40 (57.14)	80 (50.31)
HS and above	23 (25.84)	12 (17.14)	35 (22.01)
Average of Household Size	3.75	4.06	3.89
Economic status			
BPL	32 (35.96)	33 (47.14)	65 (40.88)
APL	57 (64.04)	37 (52.86)	94 (59.12)
Land holding (Acre)			
land less	55 (61.8)	26 (37.14)	81 (50.94)
<1 = Acre	32(35.96)	39(55.71)	71 (44.65)
01–02 Acre	2 (2.25)	5 (7.14)	7 (4.4)

Note Figures in the parentheses represent percentage, *Source* Field Survey, 2019

4.2.5.3 Household's Assets Structure in the Coastal Region of East Midnapore District

There are two types of assets like livestock and physical.

Livestock asset: Livestock per household in coastal region of East Midnapore district is shown in Table 4.31. It is found from Table 4.31 that people in coastal region of East Midnapore district posses mainly cow, hen, duck and goat etc. as their livestock. Average number of hen per household (2.58) is highest followed by ducks (1.84),

Table 4.30 Infrastructural facilities of sample households across villages in East Midnapore district

Infrastructure	Maitrapur (No. of HH = 89)	Purba Mukundapur (No. of HH = 70)	All (No. of HH = 159)
Sources of drinking water			
Pipe water	–	2 (2.86)	2 (1.26)
Pump well	–	6 (8.57)	6 (3.77)
river	3 (3.37)	(0)	3 (1.89)
Tube well	86 (96.63)	62 (88.57)	148 (93.08)
Sanitation facility			
No sanitation facility	22 (24.72)	12 (17.14)	34 (21.38)
Have sanitation facility	67 (75.28)	58 (82.86)	125 (78.62)
Accessibility of public health care facility			
No	14 (15.73)	10 (14.29)	24 (15.09)
Yes	75 (84.27)	60 (85.71)	135 (84.91)
Banking facility			
No	21 (23.6)	18 (25.71)	39 (24.53)
Yes	68 (76.4)	52 (74.29)	120 (75.47)
Credit facility			
No	58 (65.17)	43 (61.43)	101 (63.52)
Yes	31 (34.83)	27 (38.57)	58 (36.48)
Housing condition			
Pucca	35 (39.33)	26 (37.14)	61 (38.36)
Semi-Pucca	22 (24.72)	32 (45.71)	54 (33.96)
Katcha	32 (35.96)	12 (17.14)	44 (27.67)

Note Figures in the parentheses represent percentage, *Source* Field Survey, 2019

Table 4.31 Livestock asset per household owned in coastal region of East Midnapore district

Livestock assets per household	Maitrapur (No. of HH = 89)	Purba Mukundapur (No. of HH = 70)	All (No. of HH = 159)
Cow per household	0.70	1.01	0.84
Goat per household	0.20	1.21	0.65
hen per household	2.57	2.59	2.58
Duck per household	2.26	1.30	1.84

Note Figures in the parentheses represent percentage, *Source* Field Survey, 2019

Table 4.32 Distribution of Physical assets in coastal region of East in Midnapore district

Physical assets	Maitrapur (No. of HH = 89)	Purba Mukundapur (No. of HH = 70)	All (No. of HH = 159)
Radio	80 (89.89)	59 (84.29)	139 (87.42)
TV	41 (46.07)	52(74.29)	93(58.49)
Mobile phone	72 (80.9)	67 (95.71)	139 (87.42)
Bi-Cycle	63 (70.79)	60 (85.71)	123 (77.36)
Van	11 (12.36)	9 (12.86)	20 (12.58)
Motor bike	6 (6.74)	29 (41.43)	35 (22.01)
Refrigerator	1 (1.12)	4 (5.71)	5 (3.14)
LPG connection	24 (26.97)	37 (52.86)	61 (38.36)
Bed	72 (80.9)	62 (88.57)	134 (84.28)
Almirah	52 (58.43)	49 (70)	101 (63.52)
Fishing net	2 (2.25)	6 (8.57)	8 (5.03)
Boat	1 (1.12)	5 (7.14)	6 (3.77)

Note Figures in the parentheses represent percentage, *Source* Field Survey, 2019

goats (0.65) and cows (0.84). Number of ducks per households is higher in Maitrapur village than Purba Mukundapur village.

Physical: Physical assets in coastal region of East Midnapore consist of radio, TV and mobile, motorbike, bicycle, van, refrigerator, LPG, almirah, fishing net and boat, other furniture comprises chair, table and sofa set which are displayed in Table 4.32. It is found from this table that 87% of households have radio and mobile, 84% of households are sleeping bed, 77% of households have bicycles, 58% have TV and 38% have LPG facilities.

4.2.5.4 Livelihoods, Income and Consumption of Households of Coastal Region of East Midnapore District

There are six major sources of livelihoods which are identified in coastal region of East Midnapore and they are casual labourer, agriculture, fishing, informal employment, formal employment and van pulling activities (see Table 4.34). About 24% of household are engaged in van pulling activities and derive their income from such activities. Second source of livelihood is selling of casual labourer. About 18% households are engaged in selling of casual labour for sustenance. About 16% of households are involved in fishing activities.

Monthly income per household of different livelihood groups is shown in Table 4.35. It is observed from this table that monthly income per household in the coastal district of East Midnapore is found to Rs. 8515. Across the livelihood group the cultivating households is deriving maximum income Rs. 9531 while casual labourers are getting lowest income (Rs. 7445).

Table 4.34 Sources of livelihoods in coastal region of East Midnapore district

Major sources of livelihoods	Maitrapur	Purba Mukundapur	All
Sale of casual labour	22 (24.72)	6 (8.57)	28 (17.61)
Agriculture	10 (11.24)	10 (14.29)	20 (12.58)
Fishing	8 (8.99)	18 (25.71)	26 (16.35)
Informal employment	14 (15.73)	7 (10)	21 (13.21)
Formal employment	12 (13.48)	14 (20)	26 (16.35)
Van pulling	23 (25.84)	15 (21.43)	38 (23.9)
Total	89 (100)	70 (100)	159 (100)

Note Figures in the parentheses represent percentage, *Source* Field Survey, 2019

Table 4.35 Monthly income (Rs) per household of Livelihood groups in coastal region of Midnapore district

Livelihood groups	Maitrapur	Purba Mukundapur	All
Casual labour	7408.18	7581.67	7445.36
Cultivators	7681.63	11380.37	9531.00
Fishing communities	9638.76	7709.42	8303.07
Workers in Informal sector	8872.86	8813.00	8852.91
Workers in formal sector	6841.00	10464.19	8791.95
Van Puller	7783.39	9694.07	8537.61
All	7890.30	9309.49	8515.10

Note Figures in the parentheses represent percentage, *Source* Field Survey, 2019

Consumption: Monthly household consumption expenditure by livelihood groups in the coastal region of East Midnapore district is shown in Table 4.36. It is observed from the table that average monthly consumption expenditure per households in coastal region of East Midnapore is Rs. 4285.60. In Purba Mukundapur village consumption expenditure is higher compared to Maitrapur village. Among the livelihood groups, monthly consumption expenditure per household is highest for fishing communities (Rs. 5047.77) followed by van puller (Rs. 4463.36), cultivators (Rs. 4369.92), and workers in informal sectors (Rs. 4070.77), casual labour (Rs. 3970.30) and workers in formal sectors (Rs. 3712.45).

Table 4.36 Monthly consumption expenditure (Rs) per household of livelihood groups in Midnapore district

Livelihood groups	Maitrapur	Purba Mukundapur	All
Casual labour	3947.53	4053.8	3970.3
Cultivators	4062.4	4677.45	4369.92
Fishing communities	5673.49	4768.8	5047.17
Workers in informal sector	3935.46	4341.4	4070.77
Workers in formal sector	2966.73	4351.64	3712.45
Van Puller	4165.13	4920.64	4463.36
All	4037.67	4600.83	4285.6

Note Figures in the parentheses represent percentage, *Source* Field Survey, 2019

References

Danda AA (2007) Surviving in the Sundarbans: threats and responses. Unpublished PhD University of Twente 203

Dey S, Ghosh AK, Hazra S (2016) Review of West Bengal State adaptation policies Indian Bengal Delta (Deltas Vulnerability and Climate Change: Migration and Adaptation [DECCMA]. Working Paper). Southampton UK: DECCMA Consortium. https://generic.wordpress.soton.ac.uk/deccma/resources/working-papers/. Accessed 6 Aug 2018

Hajra R, Ghosh T (2018) Agricultural productivity household poverty and migration in the Indian Sundarban Delta. Elementa Sci Anthr 6(1):3. https://doi.org/10.1525/elementa.196

Hajra R, Szabo S, Tessler Z, Ghosh T, Matthews Z, Foufoula-Georgiou E (2017) Unravelling the association between the impact of natural hazards and household poverty: evidence from the Indian Sundarban delta. Sustain Sci 12(3):453–464

Lwasa S (2013) Planning innovation for better Urban communities in sub-Saharan Africa: the education challenge and potential responses. Town Reg Plan 60(0) (April 30):38–48

Nicholls RJ, Hanson S, Herweijer C, Patmore N, Hallegatte S, Corfee-Morlot J, Chaˆteau J, Muir-Wood R (2007) Ranking port cities with high exposure and vulnerability to climate extremes—exposure estimates. OECD environmental working paper no 1 Organisation for Economic Co-operation and Development (OECD) Paris

World Wildlife Fund (2010) Environmental management and biodiversity conservation plan for sundarbans biodiversity. Report prepared for climate change adaptation biodiversity conservation and socio-economic development of the Sundarbans Area of West Bengal World Bank non-lending technical assistance

Chapter 5
Quantitative Measurement of Vulnerability in Various Districts of West Bengal Based on Secondary Data

The present chapter attempts to measure vulnerability across 18 districts of West Bengal. The study utilizes 17 macroeconomic variables collected from secondary sources in 2001 and 2011 and estimates the vulnerability indices for all districts using exposure, sensitivity and adaptive capacity of IPCC (2007) *with equal and unequal weights. The chapter identifies more vulnerable, moderate vulnerable and less vulnerable districts of West Bengal based on vulnerability indices. Besides, the chapter also attempts to identify the determinants of vulnerability at the district level.*

5.1 District Wise Vulnerability Indices with Equal Weights and Unequal Weights

We have taken 18 sample districts of State of West Bengal like Darjeeling, Jalpaiguri, Cooch Behar, North Dinajpur, South Dinajpur, Malda, Murshidabad, Birbhum, Burdwan, Nadia, North 24 parganas, Hooghly, Bankura, Purulia, Howrah, South 24 Parganas, West Midnapore and East Midnapore.

The districts like East Midnapore, South24 Parganas, North 24 Parganas, Howrah & Hooghly are more cyclone prone areas of West Bengal. The districts like Bankura, Purulia, Birbhum and parts of Paschim Midnapore are known as drought prone areas mainly due to deficient rainfall and adverse soil structure. The landslide hazard in West Bengal has been observed mostly in the Darjeeling District and extends a part of Jalpaiguri district. The districts like Malda, Murshidabad, North and South Dinajpur, South 24 Parganas, Howrah and Hooghly are known as flood prone districts of the state. The indicators along with their functional relationship with vulnerability are described in Table 5.1.

The district wise vulnerability indices of West Bengal along with the rank of districts with equal and unequal weights in 2001 are shown in Table 5.2 and Table 5.3 respectively. We find that the vulnerability indices with equal weight are varying from

J. P. Basu, *Climate Change Vulnerability and Communities in Agro-climatic Regions of West Bengal, India*, https://doi.org/10.1007/978-3-030-50468-7_5

Table 5.1 Functional relationship with vulnerability

Components of vulnerability	Sub-indicators		Relationship with vulnerability
Exposure	E1	District wise Average Max temperature (in °C)	(+)
	E2	District wise Average Min temperature (in °C)	(+)
	E3	District wise Average Rabi season rainfall (in mm)	(+)
	E4	District wise Average Kharif season rainfall (in mm)	(+)
Sensitivity	S1	District wise Population Density (Persons per Sq. km.)	(+)
	S2	District wise Production under Total Foodgrains (Production in thousand tonnes)	(−)
	S3	District wise no. of Small farmers (in persons)	(+)
	S4	District wise Cropping intensity	(−)
	S5	District wise Main agricultural population (in person)	(+)
	S6	District wise Net Sown area (% of geographical area)	(−)
Adaptive capacity	A1	District wise Literacy rate (% of total population)	(−)
	A2	District wise total no livestock (per sq. km)	(−)
	A3	District wise % of average farm size	(−)
	A4	District wise % of PCI (at constant price 2004–2006)	(−)
	A5	District wise Length of Roads (in Kilometer)	(−)
	A6	District wise people live below poverty line (% of total population)	(+)
	A7	District wise no of people get medical facilities	(−)
		(in person)	

Table 5.2 District wise vulnerability index in 2001 with equal weights and their ranks

District	Exposure	Sensitivity	Adaptive capacity	Vulnerability
Darjeeling	0.31073 (18)	0.61446 (3)	0.38550 (18)	0.44872 (17)
Jalpaiguri	0.63368 (6)	0.47512 (10)	0.59575 (4)	0.56210 (4)
Cooch Behar	0.65692 (4)	0.38370 (13)	0.56598 (7)	0.52304 (9)
North Dinajpur	0.62302 (7)	0.30593 (17)	0.72969 (1)	0.55503 (5)
South Dinajpur	0.62302 (8)	0.35227 (14)	0.57170 (6)	0.50633 (11)
Malda	0.60961 (9)	0.34987 (15)	0.68705 (2)	0.54982 (7)
Murshidabad	0.54762 (10)	0.44512 (11)	0.51220 (13)	0.49686 (14)
Birbhum	0.54440 (11)	0.41573 (12)	0.55618 (8)	0.50384 (13)
Burdwan	0.52088 (13)	0.51395 (7)	0.44953 (16)	0.48905 (16)
Nadia	0.54002 (12)	0.26011 (18)	0.55465 (9)	0.44725 (18)
North 24 Parganas	0.66787 (3)	0.55636 (5)	0.48760 (15)	0.55428 (6)
Hoogly	0.48948 (14)	0.48445 (9)	0.53337 (10)	0.50578 (12)
Bankura	0.47802 (15)	0.55035 (6)	0.52475 (11)	0.52279 (10)
Purulia	0.45115 (16)	0.68692 (1)	0.58930 (5)	0.59125 (2)
Howrah	0.44307 (17)	0.57705 (4)	0.44039 (17)	0.48925 (15)
South 24 Parganas	0.64287 (5)	0.64757 (2)	0.52408 (12)	0.59561 (1)
West Midnapur	0.76508 (2)	0.50406 (8)	0.51120 (14)	0.56842 (3)
East Midnapur	0.78314 (1)	0.33070 (16)	0.59673 (3)	0.54670 (8)

Source author's calculation; *Note* Figures in the parentheses represent rank

0.59561 to 0.44725. The districts are ranked on the basis of vulnerability indices where rank 1 indicates most vulnerable district and rank 18 means lowest vulnerability. Considering equal weight it is indicated that the district South 24 Parganas is the most vulnerable district, its vulnerability index is highest (0.59561) and ranks first followed by the district Purulia (0.59125), the district West Midnapur (0.56842), the district Jalpaiguri (0.56210) and so on. The district Nadia is least vulnerable district of West Bengal, its rank is 18 (Table 5.2).

In terms of exposure, the district East Midnapur ranks first followed by West Midnapur, North 24 Parganas and so on. In terms of sensitivity the district of Purulia ranks first followed by the district South 24 Parganas, the district Darjeeling, the district Howrah and so on. With regard to adaptive capacity the district North Dinajpur ranks first followed by the district Malda, the district East Midnapur and so on.

By considering unequal weight, it is observed from Table 5.3 that the district South 24 Parganas is the most vulnerable district, its vulnerability index is highest (0.71499) followed by the district Jalpaiguri (0.68620), the district Purulia (0.67787) and so on. It is also found from Table 5.3 that the ranks of most of the districts in respect of exposure, sensitivity and adaptive capacity have also been changed compared with equal weight.

Table 5.3 District wise vulnerability index in 2001 with unequal weights and their ranks

District	Exposure	Sensitivity	Adaptive Capacity	Vulnerability
Darjeeling	0.08932 (18)	0.27718 (3)	0.20667 (18)	0.57319 (15)
Jalpaiguri	0.16561 (6)	0.20627 (9)	0.31431 (3)	0.68620 (2)
Cooch behar	0.16639 (5)	0.16961 (12)	0.26948 (10)	0.60549 (10)
North Dinajpur	0.14699 (7)	0.13005 (17)	0.36721 (1)	0.64426 (7)
South Dinajpur	0.14699 (8)	0.14082 (14)	0.28974 (5)	0.57756 (14)
Malda	0.14668 (9)	0.13980 (15)	0.35265 (2)	0.63914 (8)
Murshidabad	0.12397 (12)	0.19821 (11)	0.24077 (15)	0.56296 (17)
Birbhum	0.12593 (11)	0.16654 (13)	0.27372 (9)	0.56620 (16)
Burdwan	0.11677 (13)	0.23597 (7)	0.24219 (14)	0.59495 (11)
Nadia	0.13100 (10)	0.10846 (18)	0.28488 (6)	0.52436 (18)
North 24 Parganas	0.17516 (3)	0.25125 (4)	0.23901 (16)	0.66543 (5)
Hoogly	0.10964 (14)	0.20207 (10)	0.28173 (8)	0.59344 (12)
Bankura	0.10513 (16)	0.24187 (6)	0.25856 (11)	0.60557 (9)
Purulia	0.09474 (17)	0.29956 (1)	0.28357 (7)	0.67787 (3)
Howrah	0.10596 (15)	0.24968 (5)	0.22933 (17)	0.58498 (13)
South 24 Parganas	0.16912 (4)	0.29052 (2)	0.25534 (12)	0.71499 (1)
West Midnapur	0.20289 (1)	0.22016 (8)	0.25029 (13)	0.67335 (4)
East Midnapur	0.20220 (2)	0.13624 (16)	0.30984 (4)	0.64828 (6)

Source author's calculation; *Note* Figures in the parentheses represent rank

The district wise vulnerability indices in West Bengal along with ranks of the districts in 2011 with equal and unequal weights are presented in Table 5.4 and Table 5.5 respectively. It is observed from Table 5.4 that the vulnerability indices with equal weight varying from 0.68458 to 0.36211. In West Bengal, the district Purulia is the most vulnerable district on the basis of the highest value of vulnerability index (0.68458) followed by the district Bankura, the district of Jalpaiguri, the district of Howrah, the district of Darjeeling and so on. The districts like East Midnapur and West Midnapur are the least vulnerable districts on the basis of lower values of vulnerability indices (Table 5.4).

In respect to exposure, the district of Purulia ranks first followed by the district Bankura, North 24 Parganas, Jalpaiguri, and West Midnapur and so on. In terms of sensitivity, the district Purulia ranks first followed by the district Darjeeling, the district of Bankura, and the district of Howrah and so on. With regard to adaptive capacity the district Darjeeling ranks first followed by the district Purulia, the district North Dinajpur, the district Malda and so on.

From Table 5.5 it is found that the vulnerability indices varying from 0.82091 to 0.39145. It is also observed from Table 5.5 that the ranks of the districts with respect to exposure, sensitivity, adaptive capacity and vulnerability have been changed compared to the ranks with equal weights.

Table 5.4 District wise vulnerability index in 2011 with equal weights and their ranks

District	Exposure	Sensitivity	Adaptive capacity	Vulnerability
Darjeeling	0.29541 (18)	0.62601 (2)	0.72451 (1)	0.58878 (5)
Jalpaiguri	0.56620 (4)	0.59270 (5)	0.67053 (6)	0.61851 (3)
Cooch behar	0.47101 (12)	0.43671 (11)	0.59439 (10)	0.50971 (12)
North Dinajpur	0.35360 (17)	0.41327 (12)	0.70267 (3)	0.51840 (8)
South Dinajpur	0.36202 (16)	0.48939 (8)	0.62484 (8)	0.51519 (10)
Malda	0.40598 (15)	0.41003 (13)	0.68740 (4)	0.52329 (7)
Murshidabad	0.47870 (11)	0.40490 (14)	0.46314 (16)	0.44624 (14)
Birbhum	0.49466 (8)	0.49892 (7)	0.61250 (9)	0.54469 (6)
Burdwan	0.52328 (6)	0.35873 (15)	0.47537 (15)	0.44548 (15)
Nadia	0.48498 (9)	0.44339 (10)	0.59021 (11)	0.51363 (11)
North 24 Parganas	0.56765 (3)	0.52036 (6)	0.48611 (12)	0.51738 (9)
Hoogly	0.49472 (7)	0.35266 (16)	0.47783 (14)	0.43762 (16)
Bankura	0.63759 (2)	0.61507 (3)	0.68102 (5)	0.64752 (2)
Purulia	0.67675 (1)	0.65869 (1)	0.71124 (2)	0.68458 (1)
Howrah	0.45003 (14)	0.60495 (4)	0.66631 (7)	0.59377 (4)
South 24 Parganas	0.48402 (10)	0.46031 (9)	0.48139 (13)	0.47457 (13)
West Midnapur	0.54087 (5)	0.34071 (17)	0.36717 (18)	0.39870 (17)
East Midnapur	0.47058 (13)	0.27195 (18)	0.37741 (17)	0.36211 (18)

Source author's calculation; *Note* Figures in the parentheses represent rank

5.2 Comparative Analysis of District-Wise Vulnerability

The district-wise rank in vulnerability indices with equal and unequal weights in 2001 and 2011 are shown in Table 5.6. In 2001 the district South 24 Parganas was most vulnerable (rank 1) district in West Bengal but in 2011 it was less vulnerable (rank13th) district (Table 5.6). It is also observed from Table 5.6 that the district Bankura the rank in vulnerability was 10th but in 2011 its rank in vulnerability was second highest while the district Darjeeling its rank in vulnerability was 17th in 2001 but its rank in vulnerability was 5th in 2011. It also appears from Table 5.5 that the district West Midnapur was more vulnerable district, its rank in vulnerability was 3rd in 2001 while it becomes less vulnerable its rank in vulnerability was 17th in 2011. Figure 5.1 shows the position of vulnerability with equal weights from 2001 to 2011 across the districts in West Bengal. It is found from Table 5 that most of the districts ranks in vulnerability have changed. For example, the district of Bankura its rank in vulnerability was 2nd in 2011 but its rank was 10th in 2001 (Table 5.6). Similarly, the district of Darjeeling its rank in vulnerability was 4th in 2011 while its rank was 15th in 2001. Figure 5.2 shows the position of vulnerability with unequal weights from 2001 to 2011 across the districts in West Bengal.

Table 5.5 District wise vulnerability index with unequal weights and their Ranks in 2011

District	Exposure	Sensitivity	Adaptive capacity	Vulnerability
Darjeeling	0.08776 (16)	0.27455 (3)	0.33067 (3)	0.69299 (4)
Jalpaiguri	0.14120 (3)	0.26772 (4)	0.31599 (6)	0.72493 (3)
Cooch behar	0.11829 (10)	0.19739 (11)	0.28494 (10)	0.60062 (10)
North Dinajpur	0.08346 (18)	0.18393 (14)	0.34466 (1)	0.61207 (8)
South Dinajpur	0.08656 (17)	0.21722 (8)	0.29503 (8)	0.59881 (11)
Malda	0.09893 (15)	0.18493 (13)	0.33004 (4)	0.61390 (7)
Murshidabad	0.12190 (9)	0.19204 (12)	0.21928 (13)	0.53324 (14)
Birbhum	0.11324 (11)	0.22613 (7)	0.29496 (9)	0.63433 (6)
Burdwan	0.12971 (6)	0.16757 (15)	0.21372 (16)	0.51101 (15)
Nadia	0.12451 (7)	0.20581 (10)	0.27697 (11)	0.60730 (9)
North 24 Parganas	0.13982 (4)	0.23089 (6)	0.21848 (15)	0.58921 (12)
Hoogly	0.12211 (8)	0.15851 (17)	0.21862 (14)	0.49926 (16)
Bankura	0.16617 (2)	0.28440 (2)	0.32615 (5)	0.77673 (2)
Purulia	0.17839 (1)	0.29865 (1)	0.34386 (2)	0.82091 (1)
Howrah	0.11010 (14)	0.25845 (5)	0.30454 (7)	0.67314 (5)
South 24 Parganas	0.11228 (12)	0.21219 (9)	0.22148 (12)	0.54595 (13)
West Midnapur	0.13098 (5)	0.16303 (16)	0.18987 (17)	0.48390 (17)
East Midnapur	0.11170 (13)	0.11961 (18)	0.16014 (18)	0.39145 (18)

Source author's calculation; *Note* Figures in the parentheses represent rank

The distributions of vulnerable districts with equal and unequal weights are shown in Tables 5.7 and 5.8 respectively. The vulnerable district is classified into less vulnerable, vulnerable and highly vulnerable districts on the basis of the values of vulnerability indices. It is found from Table 5.7 that the number of highly vulnerable districts in West Bengal raised from 17% in 2001 to 33% in 2011 while the number of less vulnerable districts fell from 44% in 2001 to 37% in 2011 (Figs. 5.3 and 5.4).

The lists of vulnerable districts with equal weights in 2001 and in 2011 are presented in Table 5.9. The highly vulnerable districts in 2001 with equal weights are West Midnapur, Purulia and South 24 Parganas while in 2011 the districts like Darjeeling, Howrah, Jalpaiguri, Purulia and Bankura are identified as highly vulnerable districts. In the case of unequal weight there are three other districts are added like East Midnapur, North 24 Parganas and Jalpaiguri to the list of highly vulnerable district compared to equal weight in 2001. In 2011 with unequal weight the highly vulnerable districts are Bankura and Purulia. It is also observed from Table 5.8 that the less vulnerable districts with equal weight in 2001 are Nadia, Darjeeling, Burdwan, Howrah, Murshidabad, Birbhum, Hooghly and South Dinajpur while in 2011 the less vulnerable districts are East Midnapur, West Midnapur, Hooghly, Burdwan, Murshidabad, South 24 Parganas and Murshidabad. In case of unequal weight the less vulnerable districts in 2001 are Nadia and Murshidabad while in 2011 the less vulnerable

Table 5.6 Comparison of vulnerability with equal and unequal weights in 2001 and 2011

District	Vulnerability							
	Equal weight				Unequal weight			
	Index		Rank		Index		Rank	
	2001	2011	2001	2011	2001	2011	2001	2011
Darjeeling	0.448722	0.588785	17	5	0.573194	0.692995	15	4
Jalpaiguri	0.562104	0.618518	4	3	0.686203	0.724931	2	3
Cooch behar	0.523047	0.509713	9	12	0.605493	0.600626	10	10
North Dinajpur	0.555034	0.518402	5	8	0.644267	0.612074	7	8
South Dinajpur	0.506334	0.515199	11	10	0.577562	0.598818	14	11
Malda	0.549829	0.523293	7	7	0.639142	0.613909	8	7
Murshidabad	0.496864	0.446248	14	14	0.562966	0.533241	17	14
Birbhum	0.503842	0.544693	13	6	0.566201	0.634339	16	6
Burdwan	0.489058	0.445482	16	15	0.594953	0.51101	11	15
Nadia	0.44726	0.513634	18	11	0.524363	0.607306	18	9
North 24 Parganas	0.554286	0.517388	6	9	0.665436	0.589215	5	12
Hoogly	0.505781	0.437629	12	16	0.593447	0.499261	12	16
Bankura	0.522797	0.647529	10	2	0.605572	0.776735	9	2
Purulia	0.591253	0.684584	2	1	0.677878	0.820919	3	1
Howrah	0.489255	0.59377	15	4	0.584983	0.673143	13	5
South 24 Parganas	0.595616	0.474574	1	13	0.714992	0.54596	1	13
West Midnapur	0.568423	0.398703	3	17	0.673357	0.483905	4	17
East Midnapur	0.546703	0.362117	8	18	0.648284	0.39146	6	18

Source author's calculation

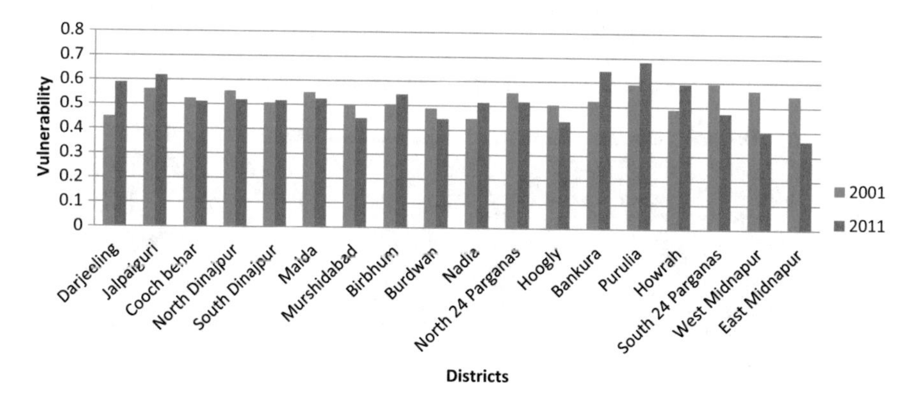

Fig. 5.1 Vulnerability indices with equal weights in 2001 and in 2011

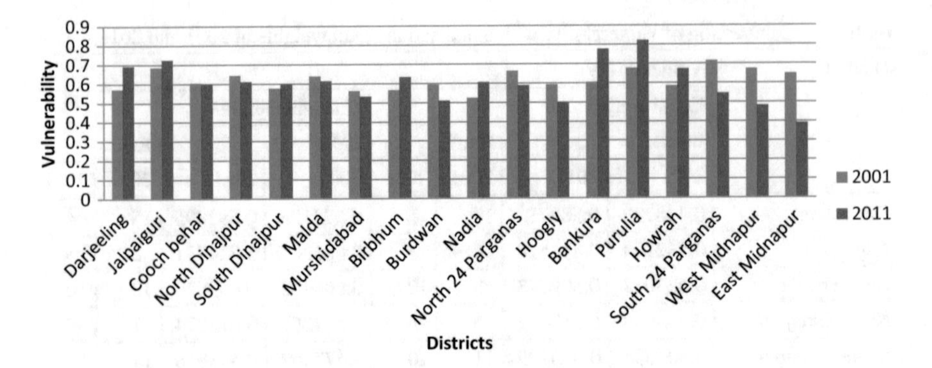

Fig. 5.2 Vulnerability indices with unequal weights in 2001 and in 2011

Table 5.7 Distribution of vulnerable districts in 2001 and 2011 with equal weights

Vulnerability index	Assigned attribute	No. of Districts	
		2001	2011
≤0.5097	Less Vulnerable	8 (44%)	7 (37%)
0.5097–0.5674	Vulnerable	7 (39%)	6 (31%)
≥0.5674	Highly Vulnerable	3 (17%)	6 (33%)

Source author's calculation; *Note* Figures in the parentheses represent percentage

Table 5.8 Distribution of vulnerable districts in 2001 and 2011 with unequal weights

Vulnerability index	Assigned attribute	No. of Districts	
		2001	2011
≤0.5629	Less Vulnerable	2 (11%)	13 (72%)
0.5629–0.6446	Vulnerable	10 (56%)	3 (17%)
≥0.6446	Highly Vulnerable	6 (33%)	2 (11%)

Source author's calculation; *Note* Figures in the parentheses represent percentage

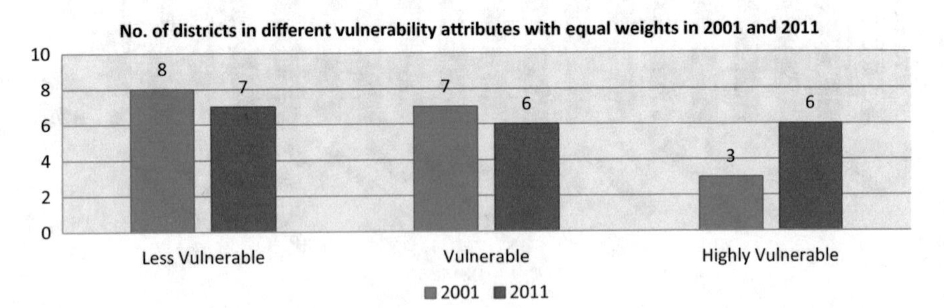

Fig. 5.3 Number of vulnerable districts in 2001 and 2011 (with equal weights)

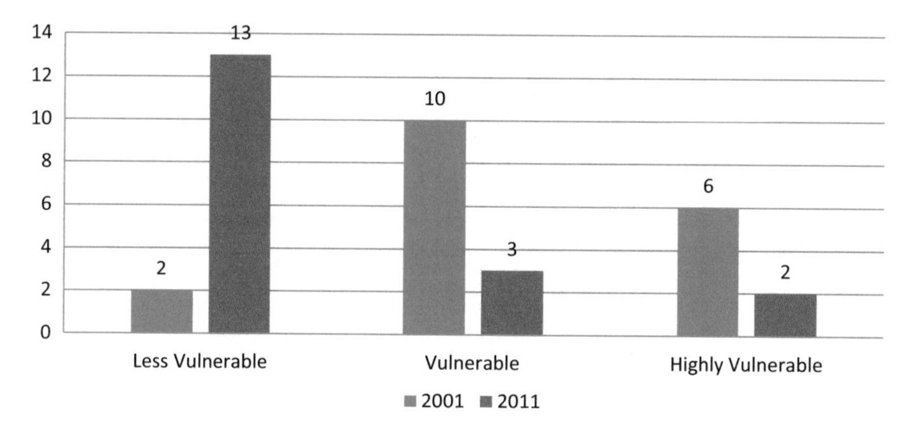

Fig. 5.4 Number of vulnerable districts in 2001 and 2011 (with unequal weights)

districts are East Midnapur, West Midnapur, Hooghly, Burdwan, Murshidabad, South 24 Parganas, Cochbehar, Nadia, South Dinajpur, North 24 Parganas, North Dinajpur, Malda and Birbhum. The lists of vulnerable districts with equal and unequal weights in 2001 and 2011 have changed (Table 5.9).

5.3 Determinants of Vulnerability by Panel Regression

We have taken 17 independent variables as described in Chap. 3, Sect. 3.3.2. The dependent variable is vulnerability indices across the districts. In order to estimate the determinants of vulnerability we apply panel regression model. We have checked the appropriateness of fixed effect model by Hausman test. The result of such test is presented in Table 5.10. This result shows that that the null hypothesis of random effect is rejected because the calculated value of Chi-square is greater than the critical values at 1% level of significance. Thus, the fixed effect model is used.

The result of fixed effect model is presented in Table 5.11. It is revealed from this table that the maximum temperature is positive and significantly related to the vulnerability. This means that the maximum temperature has an impact on vulnerability. It is also observed that the coefficients of cropping intensity and the main agricultural population are positive and significant. This shows that the cropping intensity and agricultural population are positively contributing to vulnerability. The coefficient of net sown area is positive and significant. It is also evident from Table 5.11 that the coefficient of poverty is positive and significant. This reflects that poverty has an impact on vulnerability. That is, as poverty increases the vulnerability also increases and vice versa.

Table 5.9 List of the vulnerable district in 2001 and 2011 according to their assigned attributes

Assigned attribute	Equal weight		Unequal weight	
	Identified District on 2001	Identified District on 2011	Identified District on 2001	Identified District on 2011
Less vulnerable	Nadia, Darjeeling, Burdwan, Howrah, Murshidabad, Birbhum, Hooghly, South Dinajpur	East Midnapur, West Midnapur, Hooghly, Burdwan, Murshidabad, South 24 Parganas, Murshidabad	Nadia, Murshidabad	East Midnapur, West Midnapur, Hooghly, Burdwan, Murshidabad, South 24 Parganas, Cochbehar, Nadia, South Dinajpur, North 24 Parganas, North Dinajpur, Malda, Birbhum
Vulnerable	Bankura, Coochbehar, East Midnapur, Malda, North 24 Parganas, North Dinajpur, Jalpaiguri	Nadia, South Dinajpur, North 24 Parganas, North Dinajpur, Malda, Birbhum	Birbhum, Darjeeling, South Dinajpur, Howrah, Hooghly, Burdwan, Coochbehar, Bankura, Malda, North Dinajpur	Darjeeling, Howrah, Jalpaiguri
Highly vulnerable	West Midnapur, Purulia, South 24 Parganas	Darjeeling, Howrah, Jalpaiguri, Bankura, Purulia	East Midnapur, North 24 Parganas, West Midnapur, Purulia, Jalpaiguri, South 24 Parganas	Bankura, Purulia

Source author's calculation

Table 5.10 Result of Hausman Test. Critical values Chi-square distribution at 1% level of significance is 33.40

Test Summary	Chi-square statistics	Chi-square d.f	Probability
Cross-section random	53.163295	17	0.000

Table 5.11 Results of fixed effect model

Variables	Coefficient	t-statistics	Probability
Constant	0.141401	4.066418	0.0735
Maximum temperature	0.096775	11.05103	0.05
Minimum temperature	0.044977	3.72376	0.167
Kharif season rainfall	0.012852	0.934542	0.5215
Rabi season rainfall	0.013632	0.78447	0.5765
Population density	0.042484	2.052435	0.2886
Production of total food grain	−0.010987	−0.421743	0.7459
Small farmers	−0.04589	−1.923675	0.3052
Cropping intensity	0.049148	6.79613	0.093
Main agricultural population	0.049292	7.837233	0.0808
Net sown area	0.160505	6.551703	0.0964
Literacy rate	−0.014372	−0.766029	0.5839
Livestock	0.056311	2.947193	0.2082
Farm size	0.163478	5.153422	0.122
Per capita income	−0.016185	−0.566269	0.672
Length of roads	−0.030678	−1.477212	0.3788
Poverty	0.042791	5.895481	0.098
Medical facilities	0.047285	4.975164	0.1263
Adjusted $R^2 = 0.89$; F-Statistics = 333.6			

Source author's calculation

Reference

IPCC (2007) Climate Change (2007) Impacts adaptation and vulnerability. Summary for policy makers. Intergovernmental panel on climate change (IPCC). http://www.ipcc.cg/SPMpdf

Chapter 6
Measurement of Vulnerability of Households and Determinants of Vulnerability

The present chapter attempts to measure the vulnerability of the households across five agro-climatic regions (say hill region, foot hill region, drought region and coastal region of Sundarban and coastal region of East Midnapore district) of West Bengal. In addition, this chapter classifies the degree of vulnerable households like low vulnerable, moderate and high vulnerable households across different agro climatic regions. Besides, this chapter also attempts to identify the factors responsible for vulnerability with the help order logit model of pooled data.

6.1 Vulnerability of the Households in the Hill Region of Darjeeling District

In order to calculate vulnerability index we have considered different sub-components of vulnerability along with indicators described in Sects. 3.3 and 3.3.2. The different sub-components of vulnerability are Socio-Demographic Profile (SDP), Livelihood Strategy (LS), Food, Social Network, Natural Capital, Water, Health and Climate. These 8 sub-components are taken into consideration for measuring livelihood vulnerability index at the household level. The higher values of these indices means higher vulnerability and vice versa.

The weighted livelihood vulnerability indices (LVI) along with the indices of sub-components of the vulnerability of the households in the hill region of Darjeeling district are shown in Table 6.1. The overall weighted LVI of the households in the Darjeeling district is 0.5866 while modified LVI_IPCC of the households is found to be 0.5881. It is revealed from the Table 6.1 that the Darjeeling district experiences with high food index of the sub components of LVI followed by natural capital, livelihood strategy, social network, climate variable, water, SDP and health. The high index values of food and natural capital are due to the unavailability of cultivable land and dependency of natural resources. On the other hand, low perception about

J. P. Basu, *Climate Change Vulnerability and Communities in Agro-climatic Regions of West Bengal, India*, https://doi.org/10.1007/978-3-030-50468-7_6

Table 6.1 Livelihood Vulnerability Index in the hill region of Darjeeling district

Sub component	Indicators	LVI_ Darjeeling	Weight
Socio-Demographic profile	Percentage of dependent people <20 years, >60 years	0.1298	0.3773
	Percentage of female headed households	0.0770	0.1539
	Percentage of illiteracy among households head	0.0401	0.1879
	Percentage of female family earner	0.0575	0.2809
	SDP index	**0.3043**	
Livelihood strategies	Percentage of households with family member working outside local area	0.0222	0.1191
	Percentage of households change sowing and cropping schedule	0.0458	0.0928
	Livelihood diversification index	0.1203	0.1786
	Average livestock diversification index	0.0969	0.1758
	Monthly per capita income (Rs)	0.3904	0.4336
	LS index	**0.6756**	
Food	Average crop diversity index = household has the capacity to grown at least 1 additional crop such as vegetables or pulse along with traditional crop	0.9732	1.0000
Food	**Food index**	**0.9732**	
Social network	Percentage of households not having access to communication media (like TV)	0.3455	0.4080
	Percentage of households not associated with any organization (cooperative/group) i.e. in Self Help Group (SHG)	0.1431	0.2941
	Percentage of households with non member of MGNREGA	0.1728	0.2979
	SN index	**0.6614**	
	Adaptive index	**0.6536**	
Natural capital	Percentage of households using only forest based energy for cooking	0.3228	0.5697
	Percentage of marginal and small farmer households	0.4036	0.4303
	NC index	**0.7264**	

(continued)

Table 6.1 (continued)

Sub component	Indicators	LVI_ Darjeeling	Weight
Water	Percentage of households not having regular drinking water supply	0.2886	0.3255
	Percentage of households that utilize natural water source for drinking water	0.0842	0.2106
	Percentage of households go at least 1 km to fetch water	0.0657	0.4639
	Water index	**0.4386**	
Health	Percentage of households with family member suffering from chronic illness	0.1663	0.2712
	Percentage of households not receiving treatment in local health centre	0.0729	0.3122
	Percentage of households do not have toilet facility	0.0472	0.4166
	Health index	**0.2864**	
	Sensitivity index	0.4838	
Climate variable	Percentage of households realize reduction in rainfall in past 5 years	0.1769	0.2576
	Percentage of households about the perception of lands lide increased in last 5 years	0.1665	0.2523
	Percentage of households about the perception of storm increased in last 5 years	0.1641	0.2511
	Percentage of households that did not receive warning about natural disasters	0.1195	0.2390
	CV index	**0.6270**	
	Exposure	**0.6270**	–
	LVI	**0.5866**	–
	LVI_IPCC	**0.5881**	–

Source Author's calculation from primary data

climate events and shortage of drinking water supply in hill areas explain the causes of vulnerability.

The major components of vulnerability like exposure, sensitivity and adaptive capacity indicated that there is a high adaptive index (0.6536) followed by exposure index (0.6270) and sensitivity index (0.4838). The high value of adaptive index means

Table 6.2 Classification of vulnerable households in the hill region of Darjeeling District

Assigned attribute	Vulnerability index	Number of households	%
Less vulnerable	<0.50	14	9.33
Moderate vulnerable	0.50–0.60	69	46.00
High vulnerable	>0.60	67	44.67
		150	100.00

Source Authors computation from primary data

low adaptive capacity of the households and raises vulnerability. It is revealed that high exposure also causes vulnerability of the households in the Darjeeling district.

The households are classified into three categories like less vulnerable, moderate vulnerable and high vulnerable based on vulnerability indices. Less vulnerable households are those who are in vulnerable situations but they can still cope. Moderately vulnerable households are those who are in vulnerable situations but they need urgent but temporary assistance in case of shocks and stresses while high vulnerable households are almost at a point of no return. It is found that the majority of the households (46%) are moderately vulnerable while 44.67% belong to highly vulnerable (Table 6.2). This indicates that there is a predominance of moderate and highly vulnerable households in the hill region of the Darjeeling district of West Bengal.

6.2 Vulnerability of the Households in the Foothill Region of Jalpaiguri District

The weighted livelihood vulnerability indices (LVI) along with the indices of sub-components of the vulnerability of the households in the foothill region of Jalpaiguri district are shown in Table 6.3. The overall weighted LVI and modified LVI_IPCC of the households in the Jalpaiguri district are found to be 0.5505 and 0.5343 respectively.

It is revealed from Table 6.3 that the Jalpaiguri district experiences with high value of food index followed by natural capital, livelihood strategy, social network, climate variable, water, health and SDP. High index value of food is due to incapability to produce extra crop to supplement nutrition. The higher concentration of marginal and small farmers is the main reason behind the high vulnerability of natural capital. On the other hand, low monthly per capita income and low capability in alternative diversifying income opportunities lead to the livelihood strategies more vulnerable. Again, Low perception about climatic events, low social network, low accessibility to sanitation and shortage of fresh drinking water causes for vulnerability of the households in the Jalpaiguri district.

Table 6.3 Livelihood Vulnerability Index in the foothill region of Jalpaiguri district

Sub component	Symbol of indicators	LVI_ Jalpaiguri	Weight
Socio-Demographic profile	Percentage of dependent people <20 years, >60 years	0.1390	0.3520
	Percentage of female headed households	0.0565	0.1359
	Percentage of illiteracy among households head	0.0359	0.1609
	Percentage of female family earner	0.0314	0.3511
	SDP index	**0.2628**	
Livelihood strategies	Percentage of households with family member working outside local area	0.0174	0.1506
	Percentage of households change sowing and cropping schedule	0.0504	0.0964
	Livelihood diversification index	0.2538	0.2804
	Average livestock diversification index	0.0943	0.1471
	Monthly per capita income (Rs)	0.3089	0.3255
	LS index	**0.7248**	
Food	Average crop diversity index = household has the capacity to grown at least 1 additional crop such as vegetables or pulse along with traditional crop	0.9724	1.0000
Food	**Food index**	**0.9724**	
Social network	Percentage of households not having access to communication media (like TV)	0.1131	0.3197
	Percentage of households not associated with any organization (cooperative/group) i.e. in Self Help Group (SHG)	0.1038	0.3290
	Percentage of households with non member of MGNREGA	0.2621	0.3513
	SN index	**0.4790**	
	Adaptive index	**0.6098**	–
Natural capital	Percentage of households using only forest based energy for cooking	0.2532	0.3709
	Percentage of marginal and small farmer households	0.5352	0.6291
	NC index	**0.7885**	

(continued)

Table 6.3 (continued)

Sub component	Symbol of indicators	LVI_ Jalpaiguri	Weight
Water	Percentage of households not having regular drinking water supply	0.1546	0.2093
	Percentage of households that utilize natural water source for drinking water	0.1756	0.2238
	Percentage of households go at least 1 km to fetch water	0.0611	0.5669
	Water index	**0.3913**	
Health	Percentage of households with family member suffering from chronic illness	0.0844	0.3431
	Percentage of households not receiving treatment in local health centre	0.0809	0.3507
	Percentage of households do not have toilet facility	0.1131	0.3062
	Health index	**0.2784**	**1.0000**
	Sensitivity index	**0.4861**	–
Climate variable	Percentage of households realize reduction in rainfall in past 5 years	0.0391	0.2677
	Percentage of households about the perception of flood/drought increased in last 5 years	0.2965	0.3267
	Percentage of households about the perception of storm increased in last 5 years	0.0574	0.2132
	Percentage of households that did not receive warning about natural disasters	0.1140	0.1924
	CV index	**0.5070**	
	Exposure	**0.5070**	–
	LVI_Hahn	**0.5505**	–
	LVI_IPCC	**0.5343**	–

Source Author's calculation from primary data

The households are classified into less vulnerable, moderate vulnerable and highly vulnerable on the basis of the values of Livelihood vulnerability indices (LVI). It is found that the majority of the households (36%) are moderate vulnerable while 33.08% belong to highly vulnerable and 31% of households are less vulnerable (Table 6.4). Table 6.4 indicates that there is a predominance of moderate and highly vulnerable households in the foot-hill region of the Jalpaiguri district of West Bengal.

Table 6.4 Classification of vulnerable households in the foothill regions of Jalpaiguri district

Assigned attribute	Vulnerability index	Number of HH	%
Less vulnerable	<0.50	40	30.77
Moderate vulnerable	0.50–0.60	47	36.15
High vulnerable	>0.60	43	33.08

Source Authors computation from primary data

6.3 Vulnerability of Households in the Drought Region of Purulia District

The livelihood vulnerability index of the sample households in the drought region of Purulia district is shown in Table 6.5. The overall weighted LVI of Purulia district is found to be 0.6076 while the modified LVI_IPCC is 0.6029. The overall value of LVI is explained in terms of higher values of natural capital index, of food index, of livelihood strategy index and of water index. The higher values of these indices reflect the higher vulnerability. The high dependency on natural resources for cooking and heating is the reflection of high index of natural capital. The high value of food index is due the scarcity of irrigation water for crop production. Most of the households are not taking the advantage of MGNREGA work and are unable to access sanitation facility at home.

The percentage of households fall into different categories of vulnerability is shown in Table 6.6. About 57% of the sample households belong to highly vulnerable, 29% of households belong to moderate vulnerability and only 13% of households belong to less vulnerable. The LVI indicates that the household vulnerability in Purulia district is more pronounced due to high values of exposure and sensitivity.

6.4 Vulnerability of Households in the Coastal Region of Sundarban

The overall weighted LVI of households in the coastal sunderban is found to be 0.5980 while the modified LVI_IPCC is 0.5843 (Table 6.7). It is observed from this table that the exposure index is 0.5638, the sensitivity index is found to be 0.5362 and adaptive capacity index is 0.6529.

The analysis of vulnerability also shows that the coastal people have low adaptive capacity with high exposure to climate hazards accompanied by high sensitivity. In terms of sub-components of vulnerability it is revealed that high food index, high natural capital index, and high livelihood strategy index explain the degree of vulnerability of the households in the coastal Sunderban. In addition, there exists low

Table 6.5 Livelihood Vulnerability Index in the drought region of Purulia district

Sub component	Indicators	LVI of Purulia District	Weight
Socio-Demographic profile	Percentage of dependent people <20 years, >60 years	0.1469	0.3534
	Percentage of female headed households	0.0414	0.1883
	Percentage of illiteracy among households head	0.0646	0.1588
	Percentage of female family earner	0.0847	0.2996
	SDP index	0.3375	
Livelihood strategies	Percentage of households with family member working outside local area	0.0264	0.1099
	Percentage of households change sowing and cropping schedule	0.0352	0.0978
	Livelihood diversification index	0.1286	0.2002
	Average livestock diversification index	0.1384	0.1914
	Monthly per capita income (Rs)	0.3378	0.4008
	LS index	0.6664	
Food	Average crop diversity index = household has the capacity to grown at least 1 additional crop such as vegetables or pulse along with traditional crop	0.8167	1.0000
Food	**Food index**	0.8167	
Social Network	Percentage of households not having access to communication media (like TV)	0.1370	0.3369
	Percentage of households not associated with any organization (cooperative/group) i.e. in Self Help Group (SHG)	0.1611	0.3311
		0.1793	0.3320298
	SN index	0.4774	
	Adaptive index	0.5745	

(continued)

Table 6.5 (continued)

Sub component	Indicators	LVI of Purulia District	Weight
Natural Capital	Percentage of households using only forest based energy for cooking	0.5864	0.6098
	Percentage of marginal and small farmer households	0.3251	0.3902
	NC index	0.9116	1.0000
Water	Percentage of households not having regular drinking water supply	0.2801	0.3334
	Percentage of households that utilize natural water source for drinking water	0.1678	0.2568
	Percentage of households go at least 1 km to fetch water	0.1808	0.4098
	Water index	0.6286	
Health	Percentage of households with family member suffering from chronic illness	0.1652	0.2636
	Percentage of households not receiving treatment in local health centre	0.0425	0.4251
	Percentage of households do not have toilet facility	0.2449	0.3113
	Health index	0.4526	
	Sensitivity index	0.6643	0
Climate variable	Percentage of households realize reduction in rainfall in past 5 years	0.0797	0.2298
	Percentage of households about the perception of drought increased in last 5 years	0.2710	0.3152
	Percentage of households about the perception of storm increased in last 5 years	0.0797	0.2298
	Percentage of households that did not receive warning about natural disasters	0.1397	0.2253
	CV index	0.5700	

(continued)

Table 6.5 (continued)

Sub component	Indicators	LVI of Purulia District	Weight
	Exposure	0.5700	
	LVI_Hahn	0.6076	
	LVI_IPCC	0.6029	

Source Author's calculation from primary data

Table 6.6 Classification of vulnerable households in the drought region of Purulia district

Assigned attribute	Vulnerability index	Number of households	Percentage of households
Less vulnerable	<0.50	20	13.33
Moderate vulnerable	0.50–0.60	44	29.33
High vulnerable	>0.60	86	57.33

Source Authors computation from primary data

landholding, high dependency on natural resources like fishing and crab collection and low per capita monthly income.

In the coastal region of Sunderban majority of the households (55.33%) are highly vulnerable, 23% are moderate vulnerable while 22% of households belong to less vulnerable (see Table 6.8).

6.5 Vulnerability of Households in the Coastal Region of East Midnapore District

The overall weighted LVI and modified LVI_IPCC of households in the coastal region of East Midnapore district are found to be 0.4471 and 0.3978 respectively (Table 6.9). It is also revealed from Table 6.9 that the adaptive capacity index is 0.5498, the sensitivity index is 0.3668 and the exposure is 0.2768. This shows that the low adaptive capacity is a responsible factor for explaining the vulnerability of the households in the coastal region of East Midnapore district of West Bengal.

The households are classified into less vulnerable, moderate vulnerable and highly vulnerable on the basis of the values of Livelihood vulnerability indices (LVI). It is found that the majority of the households (55%) of households are moderate vulnerable while 21% belong to highly vulnerable (see Table 6.10). The study also classifies that 24% of households belong to less vulnerable categories. Table 6.10 indicates that there is a predominance of moderate vulnerable households in the coastal region of East Midnapore district of West Bengal.

Table 6.7 Livelihood Vulnerability Index in the coastal Sunderban

Sub component	Symbol of indicators	LVI of Sunderban	Weight
Socio-Demographic profile	Percentage of dependent people < 20 years, > 60 years	0.1539	0.3848
	Percentage of female headed households	0.0543	0.1647
	Percentage of illiteracy among households head	0.0786	0.1549
	Percentage of female family earner	0.0752	0.2957
	SDP index	**0.3620**	
Livelihood strategies	Percentage of households with family member working outside local area	0.0312	0.0830
	Percentage of households change sowing and cropping schedule	0.1054	0.1207
	Livelihood diversification index	0.1230	0.2130
	Average livestock diversification index	0.0925	0.1761
	Monthly per capita income (Rs)	0.3550	0.4072
	LS index	**0.7071**	
Food	Average crop diversity index = household has the capacity to grown at least 1 additional crop such as vegetables or pulse along with traditional crop	0.9137	0.4300
Food	**Food index**	**0.9137**	
Social network	Percentage of households not having access to communication media (like TV)	0.1940	0.3266
	Percentage of households not associated with any organization (cooperative/group) i.e. in Self Help Group (SHG)	0.1981	0.3280
	Percentage of households with non member of MGNREGA	0.2367	0.3454
	SN index	**0.6288**	
	Adaptive index	**0.6529**	
Natural capital	Percentage of households using only forest based energy for cooking	0.4236	0.5282
	Percentage of marginal and small farmer households	0.3425	0.4718
	NC index	**0.7661**	

(continued)

Table 6.7 (continued)

Sub component	Symbol of indicators	LVI of Sunderban	Weight
Water	Percentage of households not having regular drinking water supply	0.1059	0.2544
	Percentage of households that utilize natural water source for drinking water	0.1015	0.2564
	Percentage of households go at least 1 km to fetch water	0.1191	0.4892
	Water index	**0.3265**	
Health	Percentage of households with family member suffering from chronic illness	0.1623	0.3330
	Percentage of households not receiving treatment in local health centre	0.1760	0.3334
	Percentage of households do not have toilet facility	0.1778	0.3336
	Health index	**0.5161**	
	Sensitivity index	**0.5362**	
Climate variable	Percentage of households realize reduction in rainfall in past 5 years	0.1511	0.2522
	Percentage of households about the perception of flood/drought increased in last 5 years	0.1390	0.2489
	Percentage of households about the perception of storm increased in last 5 years	0.1495	0.2517
	Percentage of households that did not receive warning about natural disasters	0.1242	0.2472
	CV index	0.5638	
	Exposure	0.5638	
	LVI_Hahn	0.5980	
	LVI_IPCC	0.5843	

Source Author's calculation from primary data

The level of vulnerability of the households is found to be highest in the drought region of Purulia district followed by the coastal district of Sunderban and the lowest vulnerability is observed in the East Midnapore district of West Bengal (Figs. 6.1 and 6.2).

Table 6.8 Classification of vulnerable households in the coastal region of Sunderban

Assigned attribute	Vulnerability index	Number of households	Percentage of households
Less vulnerable	<0.50	43	21.83
Moderate vulnerable	0.50–0.60	45	22.84
High vulnerable	>0.60	109	55.33

Source Authors computation from primary data

6.6 Factors Affecting the Vulnerability of the Households Using Ordered Logit Model from Pooled Data of Five Agro-Climatic Regions of West Bengal

The descriptions of the variables that affect vulnerability across five agro-climatic regions of West Bengal are presented in Table 6.11. The study categorizes household's vulnerability into three categories say less vulnerable, moderate vulnerable and high vulnerable. The households whose livelihood vulnerability index less than 0.50 are treated as low vulnerable, those households have livelihood vulnerability indices more than 0.50 but less than 0.60 are treated as moderate vulnerable and those households have livelihood vulnerability indices greater than 60 are known as high vulnerable.

The present study utilizes order logit model to identify the factors that determine the level of vulnerability of a household. To do this the study arranges the households according to their degree of vulnerability. Low vulnerable households are assigned to 1, moderate vulnerable households are assigned to 2 and high vulnerable households are assigned to 3. Hence, the dependent variable is in ascending order (1, 2 and 3) of households according to their livelihood vulnerability indices. Descriptions of 16 socio-economic-demographic-climatic factors that are considered as explanatory variables of vulnerability along with their expected relationship with vulnerability are displayed in Table 6.11. The STATA 12.0 version is used to diagnose the order logistic model.

The results of order logit model on pooled data of five districts are presented in Table 6.12. The estimated McFadden's Pseudo-R^2 explains the goodness of fit of the logit model. The estimated Pseudo-R^2 s in the present model is 0.3221. On the basis of the values of LR Chi-square, log likelihood and P-value the model is overall statistically significant at 1% level.

The result shows that vulnerability due to climate changes across districts is statistically significant except Darjeeling and East Midnapore district. This implies that the households in Purulia, Jalpaiguri and Sunderban (south 24 Parganas district) are more vulnerable compared to the district of Darjeeling and the district of East Midnapore. Of the explanatory variables, the coefficients of seven variables are negative and significant. These explanatory variables are number of adaptation strategies, caste,

Table 6.9 Livelihood vulnerability index of the households in the coastal region of East Midnapore district

Sub component	Indicators	LVI of East Midnapore district	Weight
Socio-Demographic profile	Percentage of dependent people <20 years, >60 years	0.1318	0.2743
	Percentage of female headed households	0.0291	0.1925
	Percentage of illiteracy among households head	0.0197	0.2609
	Percentage of female family earner	0.0385	0.2722
	SDP index	0.2191	
Livelihood strategies	Percentage of households with family member working outside local area	0.0137	0.1678
	Percentage of households change sowing and cropping schedule	0.0334	0.0967
	Livelihood diversification index	0.1776	0.2862
	Average livestock diversification index	0.1215	0.2309
	Monthly per capita income (Rs)	0.1625	0.2184
	LS index	0.5088	
Food	Average crop diversity index = household has the capacity to grown at least 1 additional crop such as vegetables or pulse along with traditional crop	0.9673	
Food	**Food index**	0.9673	
Social network	Percentage of households not having access to communication media (like TV)	0.1357	0.3270
	Percentage of households not associated with any organization (cooperative/group) i.e. in Self Help Group (SHG)	0.1340	0.3278
	Percentage of households with non member of MGNREGA	0.2345	0.3452
	SN index	0.5042	

(continued)

Table 6.9 (continued)

Sub component	Indicators	LVI of East Midnapore district	Weight
	Adaptive index	0.5498	
Natural capital	Percentage of households using only forest based energy for cooking	0.1574	0.2701
	Percentage of marginal and small farmer households	0.6696	0.7299
	NC index	0.8270	
Water	Percentage of households not having regular drinking water supply	0.0668	0.1800
	Percentage of households that utilize natural water source for drinking water	0.0213	0.3762
	Percentage of households go at least 1 km to fetch water	0.0393	0.4439
	Water index	0.1274	
Health	Percentage of households with family member suffering from chronic illness	0.0373	0.3954
	Percentage of households not receiving treatment in local health centre	0.0487	0.3228
	Percentage of households do not have toilet facility	0.0603	0.2818
	Health index	0.1463	
	Sensitivity index	0.3669	
Climate variable	Percentage of households realize reduction in rainfall in past 5 years	0.0334	0.3122
	Percentage of households about the perception of drought increased in last 5 years	0.0653	0.2078
	Percentage of households about the perception of storm increased in last 5 years	0.0397	0.2743
	Percentage of households that did not receive warning about natural disasters	0.1384	0.2057
	CV index	0.2768	

(continued)

Table 6.9 (continued)

Sub component	Indicators	LVI of East Midnapore district	Weight
	Exposure	0.2768	–
	LVI_Hahn	0.4471	–
	LVI_IPCC	0.3978	–

Table 6.10 Classification of vulnerable households in the coastal region of East Midnapore district

Assigned attribute	Vulnerability index	Number of households	Percentage of households
Less vulnerable	<0.50	38	23.90
Moderate vulnerable	0.50–0.60	87	54.72
High vulnerable	>0.060	34	21.38

Source Authors computation from primary data

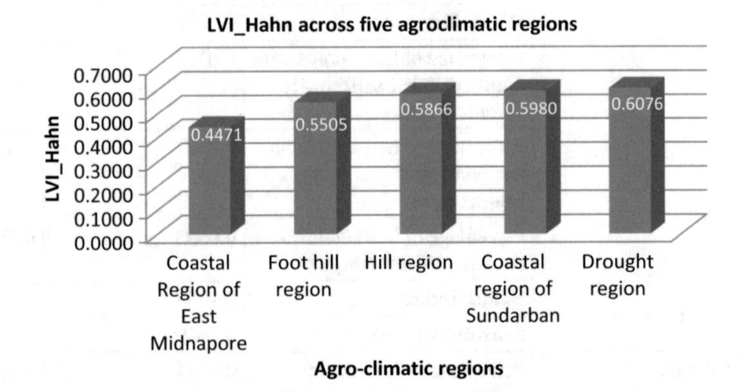

Fig. 6.1 Household's vulnerability (LVI_Hahn) in five agro-climatic regions

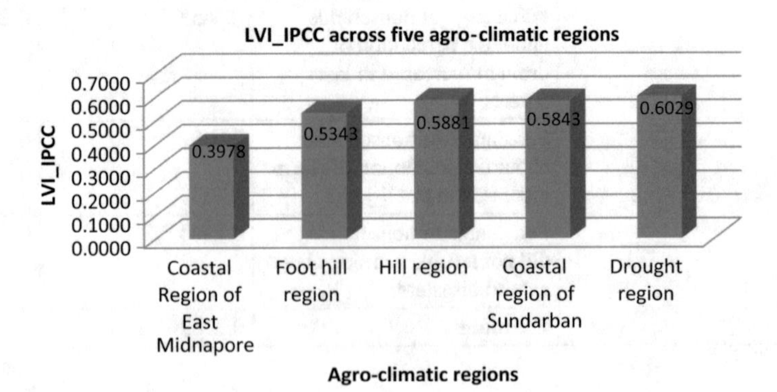

Fig. 6.2 Household's vulnerability (LVI_IPCC) in five agro-climatic regions

Table 6.11 Description of independent variables of vulnerability

Independent variables	Description of variables that explains vulnerability	Unit	Expected relation with vulnerability
Agro-climatic region dummy	D1 = 1 when hill region of Darjeeling district D1 = 0 otherwise D2 = 1 when foot-hill region of Jalpaiguri district D2 = 0, otherwise. D3 = 1 when drought prone region of Purulia district D3 = 0, otherwise. D4 = 1 when coastal region of Sundarban D4 = 0, otherwise. D5 = 1 when coastal region of East Midnapore district D5 = 0, otherwise		
Adaptation	Number of Adaptation strategies	Number	–
Household size	Total family member of the household	Person	+
Caste	SC = 1,ST = 2, OBC = 3, GEN = 4		–
Age	Age of head of the family	Years	+
Gender	Sex of the household head	Male = 1, female = 0	–
Edu_ Years	Education of Head of the Households by Years of schooling	Years	–
Marital status	Whether the head of the family married or unmarried	Married = 1, Unmarried = 0	
Land holdings	Operational holding = own land + leased in land- leased out land	acre	–
Male income earner	Total number of adult earning male in the household	Person	–
Information	Access to information		–
Physical Asset value	Total value of the physical assets of the households	Rs	–

(continued)

Table 6.11 (continued)

Independent variables	Description of variables that explains vulnerability	Unit	Expected relation with vulnerability
Off farm income	Percentage share of total income from wage lab, forestry, fishing, crab collection and formal sources		–
Farm Income	Percentage share of total income from crop and livestock	Rs	+
Non-farm income	Percentage share of total income from trading, transport and informal(worker in shop)	Rs	–
Perception index on climate variables	Average of normalized score of the climate related variables, increase in hot days, increase in cold days, increase in rainy days, Unusual formation of fog, increase in stormy events etc		–
Dependent variables	Scale of vulnerability	Order	
Low vulnerable	<0.5	1	
Moderate vulnerable	0.51–0.60	2	
High vulnerable	>0.60	3	

gender, education, operational holding, access to climate information and non-farm income. Only one variable i.e. off-farm income is positively and significantly related to vulnerability.

Number of adaptation strategies and vulnerability is negatively related to each other. That means if a household undertakes larger numbers of adaptation strategies, there will be lower vulnerability. There is a negative relationship between caste and vulnerability. This implies that lower caste households (such as SC and ST) are more vulnerable compared to higher caste (OBC and General) households. Gender is also inversely related to the vulnerability of households. It implies that female headed households are more vulnerable compared to male headed households. There exists also a negative relation between education and vulnerability. It reveals that higher the education level of household head lower will be the vulnerability of that household. Possession of cultivable land is an inversely related with vulnerability. If a household possesses more cultivable land the households will face lower vulnerability compared to the households with less amount of land possession. Accessibility of climatic information and vulnerability is negatively related. This means that higher

Table 6.12 Result of ordered logit model for identifying factors of vulnerability of pooled data of five sample districts

	Coef.	Std. Err.	z	P > |z|	[95% Conf.Interval]	
Purulia	−1.386	0.362	−3.83	0.000	−2.0966	−0.67623
Jalpaiguri	−1.257	0.311	−4.04	0.000	−1.86636	−0.64711
Darjeeling	0.036	0.260	0.14	0.891	−0.47464	0.545957
South24parganas	−0.466	0.091	−5.11	0.000	−0.645	−0.287
East Midnapore	−0.383	0.266	−1.44	0.149	−0.9041	0.13742
No. of adaptation strategies	−0.531	0.073	−7.33	0.000	−0.67359	−0.38935
Family size	−0.020	0.053	−0.39	0.698	−0.1236	0.082776
caste	−0.199	0.081	−2.47	0.013	0.041374	0.357141
Age (years)	−0.007	0.005	−1.38	0.166	−0.0179	0.003084
sex	−0.637	0.162	−3.93	0.000	− 0.95449	−0.3191
Education (years)	−0.096	0.018	−5.41	0.000	−0.13021	−0.06096
Marital status	0.243	0.301	0.81	0.419	−0.34645	0.832989
Operational holdings	−0.310	0.127	−2.44	0.015	−0.55899	−0.06071
Access to climatic information	−0.764	0.171	−4.46	0.000	−1.09926	−0.4281
Total no. of male earners	−0.185	0.119	−1.55	0.121	−0.41863	0.048736
Physical asset	0.000	0.000	1.35	0.178	−2.19E-07	1.18E-06
Farm income	0.013	0.008	1.52	0.129	−0.00371	0.029223
Off farm income	0.017	0.007	2.34	0.019	0.002839	0.031956
Non-farm income	−0.014	0.007	−1.93	0.054	−0.00022	0.02844
Perception index	−0.389	0.363	−1.07	0.284	−1.09942	0.322299
Number of observation = 786; LR $chi^2(18) = 202.05$; Prob > $chi^2 = 0.0000$						
Log likelihood = −726.36144 Pseudo $R^2 = 0.3221$						

the accessibility of climatic information lower will be the vulnerability and vice versa. Non-farm income and vulnerability is also inversely related. It implies that higher the percentage share of income of a household comes from non-farm sector then the household will face lower vulnerability. Off-farm income and vulnerability is positively related. It implies that if percentage share of off-farm income to total income of a household is higher, vulnerability of that household will also be higher.

Chapter 7
Measurement of Vulnerability of Occupational Groups of Households in Different Agro Climatic Regions of West Bengal

This chapter measures vulnerability of different occupational groups of households across five agro-climatic regions of West Bengal. The Livelihood Vulnerability Index (LVI) of Hahn et al. (2009) *and LVI_IPCC Index are used to measure the vulnerability of different occupational groups. The details of the methods, indicators, sub components and major components adapted for different occupational group wise vulnerability indices are described in Chap.* 3.

7.1 Vulnerability of Different Occupational Groups of Households in the Hill Region of Darjeeling District

The sample households of Darjeeling district are classified by major occupational groups into tea garden labour, tourist guide cum driver, workers in the informal sector, petty businessmen, casual labour, and workers in the formal sector. The distribution of occupational group of households in the hill region of Darjeeling district and their descriptions are presented in Table 7.1. It is revealed that 7% of sample households belong to tea garden labour, 28% households belong to tourist guide cum driver, 6% households belong to informal sectors, 19% households belong to petty business, 9% households belong to casual labour and 31% households belong to the formal sector (see Table 7.1 and Fig. 7.1).

The livelihood vulnerability indices for different occupational group of households in the district of Darjeeling are presented in Table 7.2. The livelihood vulnerability indices of different occupational groups vary from 0.5339 to 0.6275. It is observed from Table 7.2 that the tea garden labourers in the Darjeeling district are most vulnerable (0.6275) followed by the casual labourers (0.5830), the workers in the informal sector (0.5796), the petty businessmen (0.5732), tourist guide cum drivers (0.5697) and workers in the formal sector (0.5339). It is revealed that the workers in formal sector are the least vulnerable occupational group (0.5339) in the

J. P. Basu, *Climate Change Vulnerability and Communities in Agro-climatic Regions of West Bengal, India*, https://doi.org/10.1007/978-3-030-50468-7_7

Table 7.1 Distribution of different occupational group of households in the Hill region of Darjeeling District

Occupational groups	Description	Number of households (%)
Tea garden labour households (N_1)	Households who are working in tea garden only	10 (7)
Tourist guide and driving (N_2)	Households who are engaged as tourist guide and driving	42 (28)
Workers in the informal sector (N_3)	Households who are working in hotel, restaurant, shop, shopping mall and factory etc.	9 (6)
Petty Business men (N_4)	Households who are engaged in business like tea stall, vending, grocery etc.	28 (19)
Casual labour (N_5)	Households who are engaged in construction of road, building, drainage etc.	14 (9)
Formal sector (N_6)	Employment in the government sector like army and banking	47 (31)
All ($N_1 + N_2 + N_3 + N_4 + N_5 + N_6$)		**150 (100)**

Source Computed by author from field survey; Figures in the parenthesis represent percentage

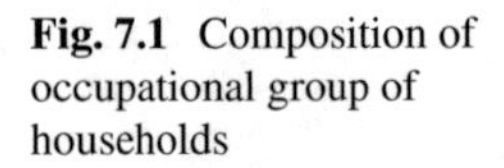
Fig. 7.1 Composition of occupational group of households

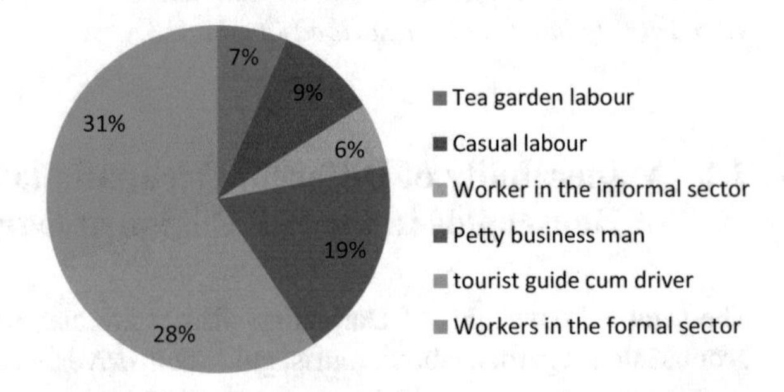

Darjeeling district (Table 7.2). Figure 7.2 shows the diagrammatic representation of vulnerability indices of different occupational group of households in the Darjeeling district.

The nature of vulnerabilities is described in terms of different sub-component of vulnerability. The different sub-components of vulnerability are socio-demographic profile (SDP), Livelihood strategy (LS), Food, Social Network, Natural capital, Water, Health and Climate. These sub-components are taken into consideration for measuring livelihood vulnerability index of different occupational groups. The higher values of these indices show higher vulnerability and vice versa. The tea garden labourers are more vulnerable than other occupational groups due to higher values of food index, natural capital index and climate index (Table 7.2). The high dependency of natural capital, inaccessibility to food and less perception about climatic events are responsible for higher vulnerability. The high value of food index is due to low value of crop diversity (it is <1). It means that the dependency of food crop is very

Table 7.2 Indices of sub components and vulnerability indices for different occupational group of households in the hill region of Darjeeling district, West Bengal

Sub-components of vulnerability	Tourist guide cum driver $N_1 = 42$	Petty business men $N_2 = 28$	Workers in informal sectors $N_3 = 9$	Workers in formal sectors $N_4 = 47$	Casual labour $N_5 = 14$	Tea garden labour $N_6 = 10$
Socio-demographic profile index	0.2739	0.3173	0.3644	0.2494	0.3705	0.4498
Livelihood strategy index	0.7294	0.7461	0.8533	0.6048	0.7296	0.4460
Food index	0.9774	0.9352	0.9944	0.9795	0.9911	0.9860
Social network index	0.7462	0.4571	0.5952	0.6117	0.4430	0.4738
Natural capital index	0.7140	0.6215	0.4551	0.6274	0.6488	0.9669
Water index	0.3234	0.6224	0.3814	0.3288	0.3948	0.4001
Health index	0.2269	0.2933	0.3452	0.2334	0.3190	0.5467
Climate variable index	0.5664	0.5925	0.6476	0.6361	0.7673	0.7506
Livelihood vulnerability index (by Hahn et al. 2009)	**0.5697 (5)**	**0.5732 (4)**	**0.5796 (3)**	**0.5339 (6)**	**0.5830 (2)**	**0.6275 (1)**

Source Computed by author from primary data; Figures in the parenthesis represent rank

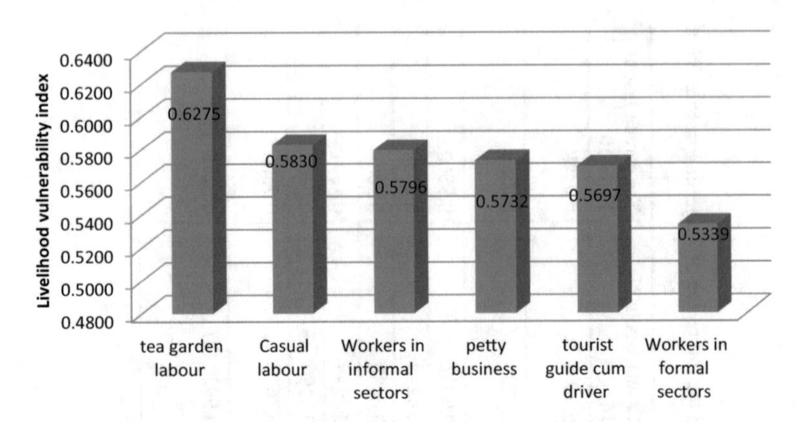

Fig. 7.2 LVI of different occupational groups in Hill district of Darjeeling, West Bengal

high in the hill region of Darjeeling district. The workers in the informal sector and petty businessmen are less affected by socio-demographic variable, natural capital health and water (Table 7.2). Again, low vulnerability of the workers in the formal sector is due to that fact that it is less affected by socio-demographic variables, natural capital, water, health and climate variable. The livelihood vulnerability of different sub-components for different occupation groups are shown in the Rader diagram (Fig. 7.3).

Intergovernmental Panel on Climate Change (IPCC 2007) defined vulnerability is the functions of contributory factors like exposure, sensitivity and adaptive capacity. Adaptive capacity comprises the weighted average of Socio Demographic Profile (SDP), Livelihood Strategies (LS), Food and Social Network (SN). The sensitivity index is measured by the weighted averages of Natural Capital, Water and Health while exposure index is determined by climate indicators. The modified livelihood vulnerability indices (LVI_IPCC) for different occupational group of households in Darjeeling district are shown in Table 7.3. It is observed from Table 7.3 that vulnerability of tea garden labourers is highest (0.6591) and lowest for workers in the informal sector (0.5480). The more vulnerable communities are tea garden labourers followed by casual labourers, workers in the informal sector, petty businessmen, tourist cum driver etc. in the hill region of Darjeeling district. The tea garden labourers are facing with higher exposure and sensitivity compared to the other occupational group of households (Table 7.3). This Table shows that the adaptive capacity is lowest for tea garden labourers and highest for the workers in the informal sector. Figure 7.4 shows the diagrammatic representation of modified livelihood vulnerability indices for different occupational group of households in the Darjeeling district. The Rader diagram for modified livelihood vulnerability indices of different sub-components for different occupational group of households are shown in Fig. 7.5.

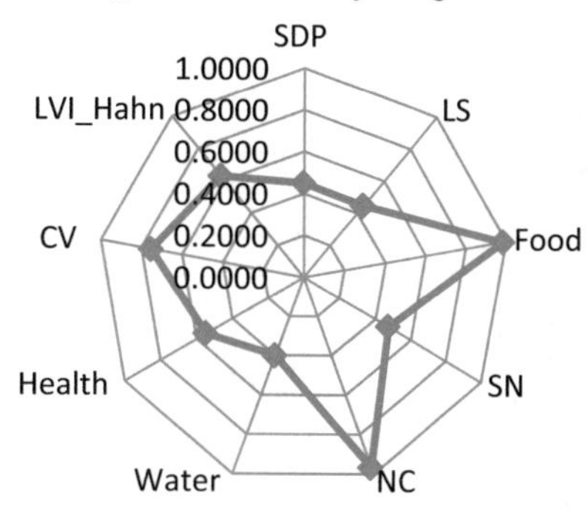

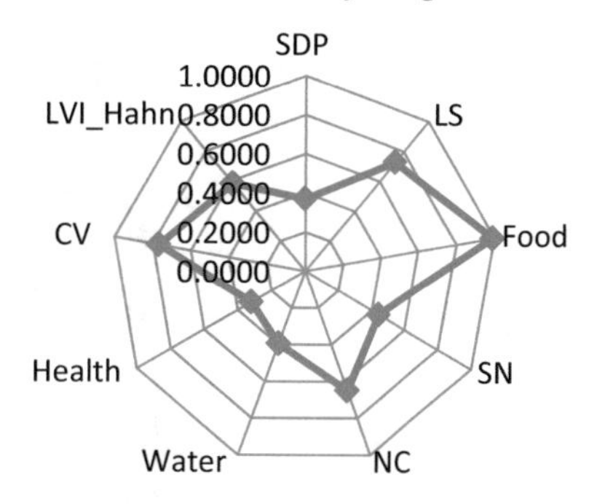

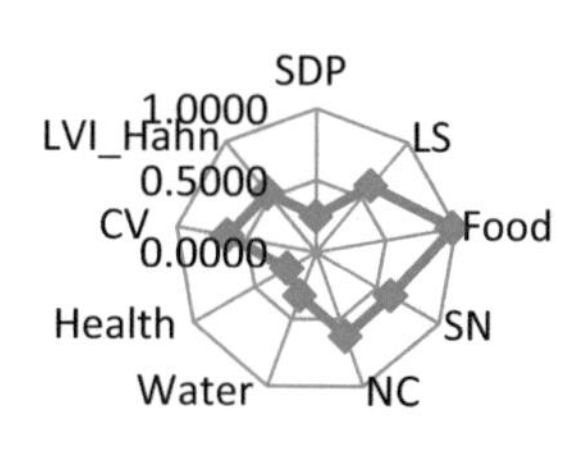

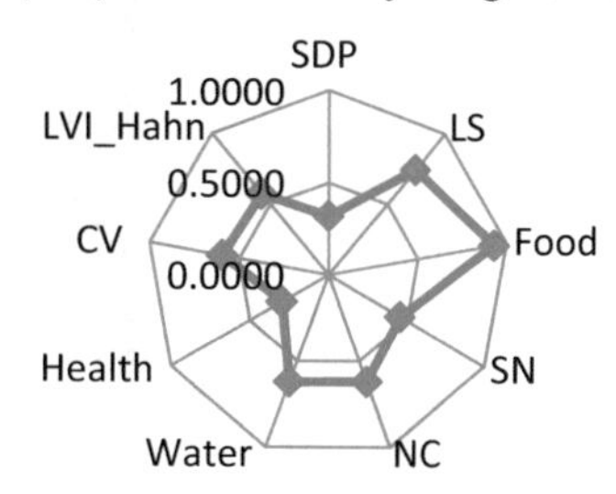

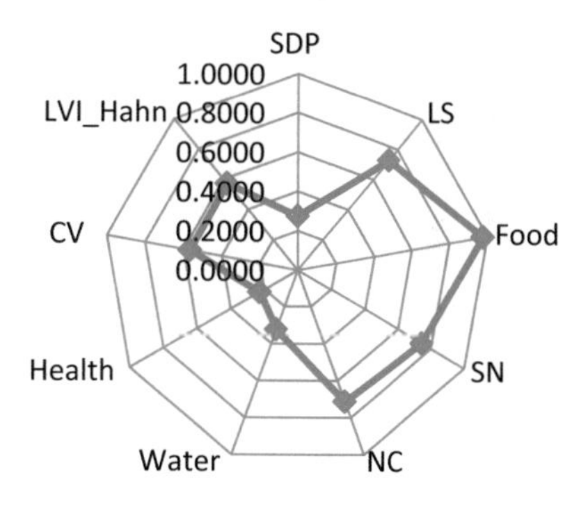

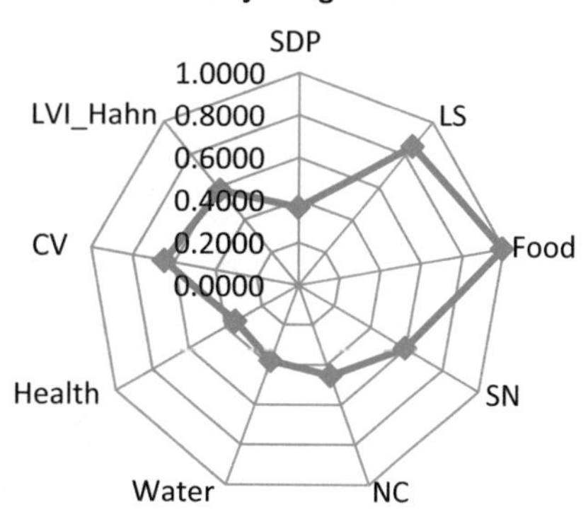

Fig. 7.3 Rader diagram for LVI_Hahn for different occupation groups in the Darjeeling district of West Bengal

Table 7.3 Modified Vulnerability indices (LVI_IPCC) for different occupational group in hill region of Darjeeling district, West Bengal

Components of vulnerability	Tourist guide cum driver $N_1 = 42$	Petty business men $N_2 = 28$	Workers in informal sectors $N_3 = 9$	Workers in formal sectors $N_4 = 47$	Casual labour $N_5 = 14$	Tea garden labour $N_6 = 10$
Exposure index	0.5664	0.5925	0.6476	0.6361	0.7673	0.7506
Sensitivity index	0.4214	0.5124	0.3939	0.3965	0.4542	0.6379
Adaptive index	0.6817	0.6139	0.7018	0.6114	0.6335	0.5889
LVI_IPCC	**0.5565 (5)**	**0.5729 (4)**	**0.5811 (3)**	**0.5480 (6)**	**0.6184 (2)**	**0.6591 (1)**

Source Computed by author from primary data; Figures in the parenthesis represent rank

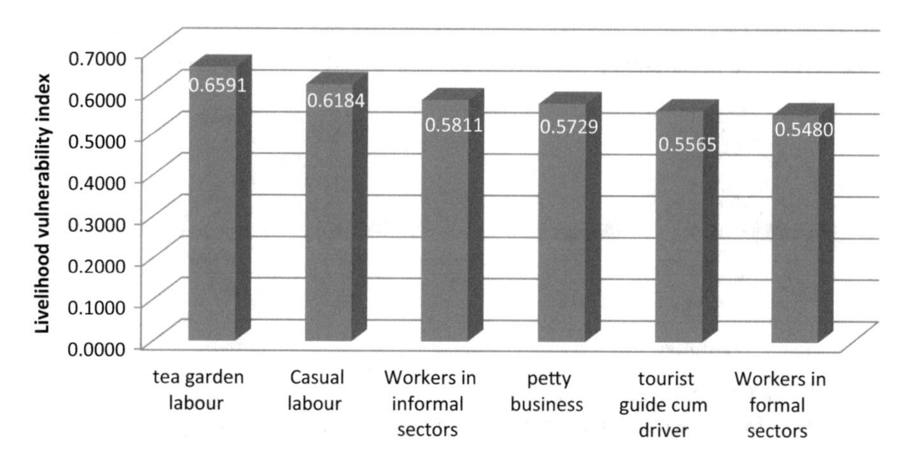

Fig. 7.4 Modified Vulnerability indices (LVI_IPCC) for livelihood groups in Hill region of Darjeeling district

7.2 Vulnerability of Different Occupational Group of Households in the Foothill Region of Jalpaiguri District

The sample households of Jalpaiguri district are classified by major occupational groups into cultivators, forest dependent communities, workers in the informal sector, casual labour, and workers in the formal sector. The distribution of different occupational group of households and their descriptions are presented in Table 7.4. It is revealed that 14% sample households belong to cultivators, 5% as forest dependent communities, and 58% belong to informal sectors. 13% to casual labour and 10% to the formal sector (see Table 7.4 and Fig. 7.6).

The livelihood vulnerability index of different occupational group of households in Jalpaiguri district is shown in Table 7.5. The vulnerability indices vary from 0.4594 to 0.5809. In the Table 7.5 it is revealed that the casual labourer are most vulnerable group in the district of Jalpaiguri with index value 0.5809 followed by forest dependent community (0.5755), workers in the informal sectors (0.5754), cultivators (0.5446) and workers in the formal Sectors (0.4594). The vulnerability of the workers in the formal sector is found to be lowest (see Fig. 7.7).

In terms of different sub-component of vulnerability, the casual labourers are more vulnerable than other occupational groups due to higher values of livelihood strategy index, food index, natural capital index and water index (Table 7.5). On the other hand, the workers in the formal sector are less affected by natural capital, social network. Again, the forest dependent communities are affected much by livelihood strategy index, food index, natural capital index and climate variable index. The sub components of vulnerability indices for different occupational group of households are shown in Fig. 7.8.

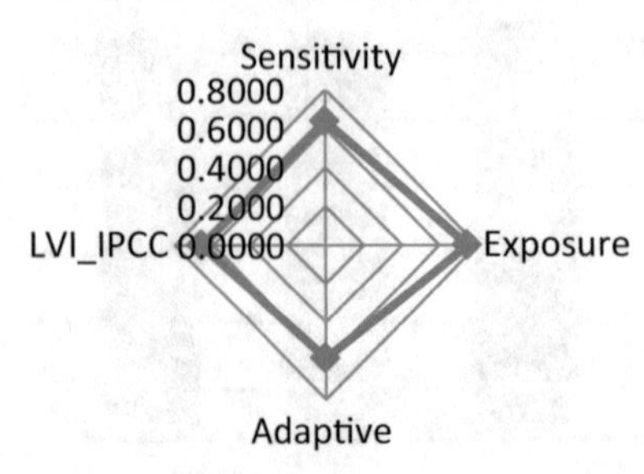

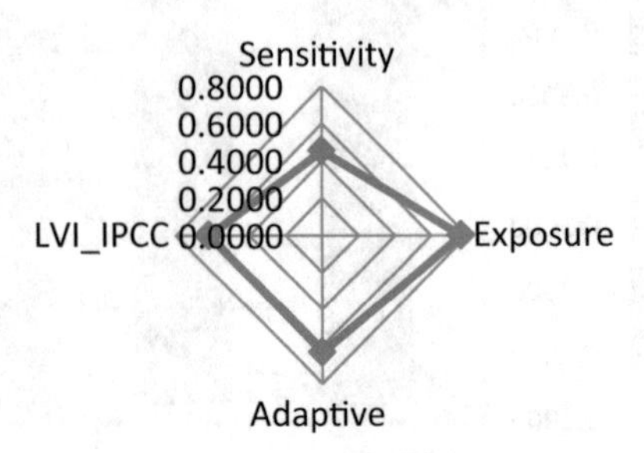

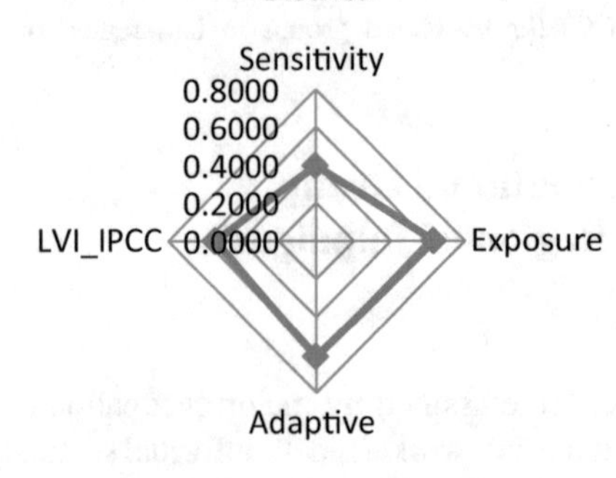

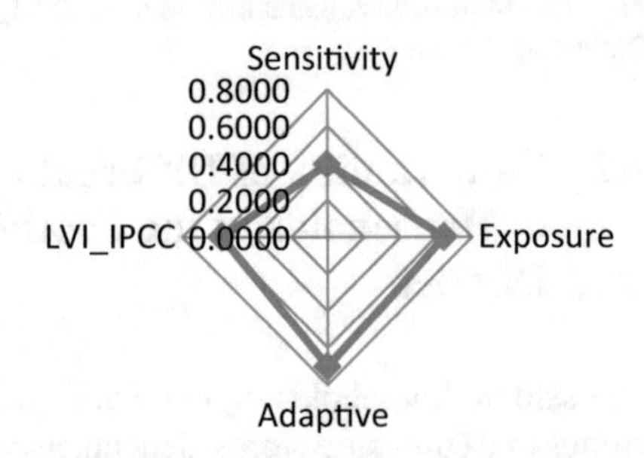

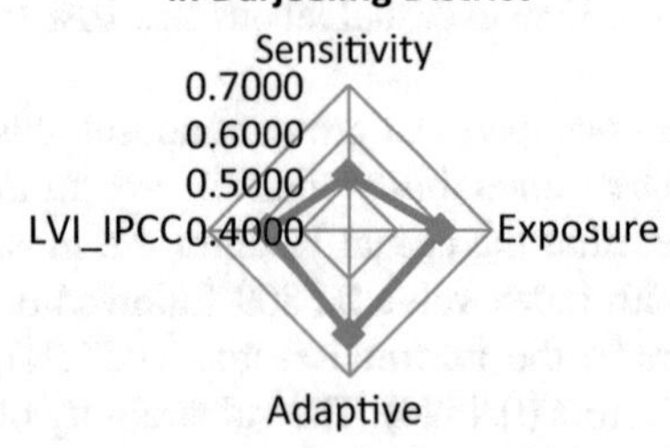

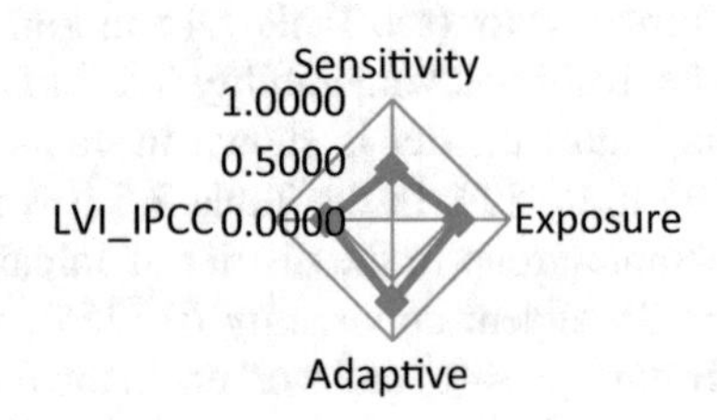

Fig. 7.5 Rader diagram for LVI_IPCC for different occupation groups in hill region of Darjeeling district

Table 7.4 Distribution of different occupational group of households in the Foothill region of Jalpaiguri district

Occupational group of households	Description	Number of households (%)
Cultivators (N_1)	Households whose main occupation is cultivation of agricultural crops	19 (14)
Forest dependent communities (N_2)	Households who are dependent on forests for livelihood generation	6 (5)
Workers in the informal sector (N_3)	Households who are working in hotel, restaurant, shop and factory etc.	75 (58)
Workers in the formal sector (N_4)	Households who are employed in government services	13 (10)
Casual labour (N_5)	Households who are engaged in construction of road, building, drainage etc.	17 (13)
All ($N_1 + N_2 + N_3 + N_4 + N_5$)		**130 (100)**

Source Computed by author from field survey; Figures in the parenthesis represent percentage

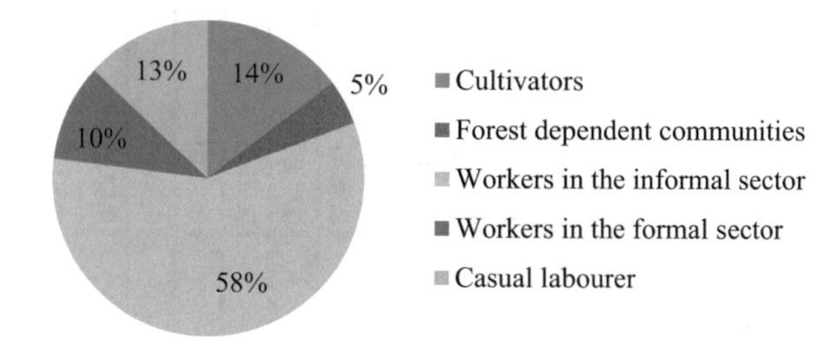

Fig. 7.6 Composition of occupational group of households in Jalpaiguri district

The modified livelihood vulnerability indices (LVI_IPCC) for different occupational group of households in Jalpaiguri district are shown in Table 7.6. It is observed from Table 7.6 that vulnerability of the forest dependent communities are found to be highest (0.5652) and lowest for workers in the formal sector (0.4111). The forest dependent communities are facing with high exposure and low adaptive capacity compared to the other occupational group of households. Figure 7.9 shows the diagrammatic representation of modified livelihood vulnerability indices for different occupational group of households in the district of Jalpaiguri. The Rader diagram for modified livelihood vulnerability indices of different sub-components of different occupational group of households are shown in Fig. 7.10.

Table 7.5 Indices of sub components and vulnerability indices for different occupational group in foothill region of Jalpaiguri district, West Bengal

Sub-components of vulnerability	Cultivators $N_1 = 19$	Forest dependent communities $N_2 = 6$	Workers in informal Sectors $N_3 = 75$	Workers in formal sectors $N_4 = 13$	Casual labour $N_5 = 17$
Socio-demographic profile index	0.2420	0.3663	0.2417	0.2985	0.2909
Livelihood strategy index	0.9572	0.9832	0.9687	0.7200	0.9619
Food index	0.9342	0.8194	0.9489	0.9040	0.9804
Social network index	0.3985	0.6111	0.5083	0.4103	0.4298
Natural capital index	0.6228	0.7409	0.7420	0.4886	0.7207
Water index	0.3742	0.3270	0.3515	0.3045	0.4012
Health index	0.1825	0.1667	0.3381	0.2451	0.3731
Climate variable index	0.6455	0.5890	0.5039	0.3041	0.4895
Livelihood vulnerability index (by Hahn et al. 2009)	**0.5446 (4)**	**0.5755 (2)**	**0.5754 (3)**	**0.4594 (5)**	**0.5809 (1)**

Source Computed by author from primary data; Figures in the parenthesis represent rank

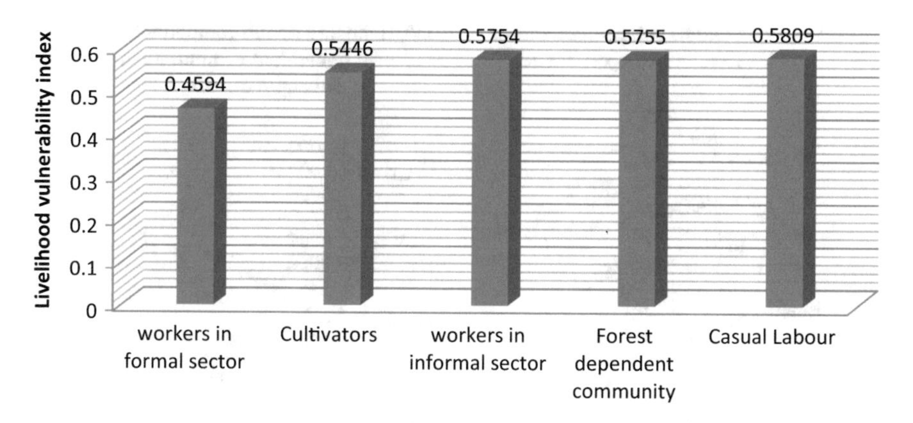

Fig. 7.7 LVI of different occupational group of households in the Foot hill district of Jalpaiguri, West Bengal

7.3 Vulnerability of Different Occupational Group of Households in the Drought Prone District of Purulia

In terms of major occupational groups sample households of Purulia district are classified as cultivators, forest dependent communities, workers in the informal sector and casual labourers. The distribution of sample households by occupational group and their descriptions are presented in Table 7.7. It is revealed from this Table 7.7 that 33% sample households belong to cultivators, 26% as forest dependent communities, 28% belong to informal sector and 13% belong to casual labourers (see Table 7.7 and Fig. 7.11).

The livelihood vulnerability indices for different occupational group of households in the district of Purulia are shown in Table 7.8. This table reveals that the casual labourer are most vulnerable with index value 0.6426 followed by forest dependent communities (0.6248), cultivators (0.6109) and worker in informal sector (0.5814). It is also found from Table 7.8 that the workers in the informal sector are least vulnerable.

Figure 7.12 shows the diagrammatic representation of vulnerability indices for different occupational group of households in the district of Purulia. The nature of vulnerabilities is described in terms of different sub-components of vulnerability. The higher values of these indices means higher vulnerability and vice versa. Casual labourers are more vulnerable than other occupational groups due to higher values of socio-demographic profile index, higher values of livelihood strategy index, social network index, and natural capital index (Table 7.8). This means that the vulnerability of casual labourers is explained by poor socio-economic conditions, low access to information and communication technology (ICT), not associated with the member of SHGs, exclusion from government beneficiaries list of MGNREGA, lower land holdings and more dependent on natural capital. The vulnerability of forest dependent communities is explained by higher values of food index, higher values of natural

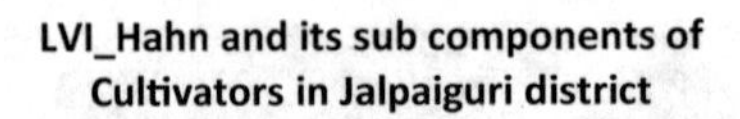

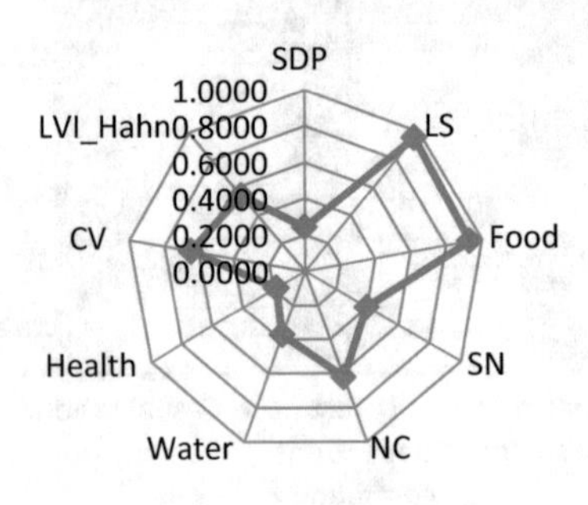

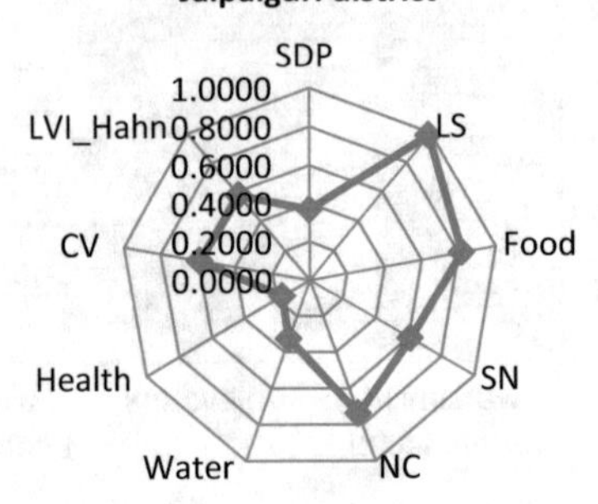

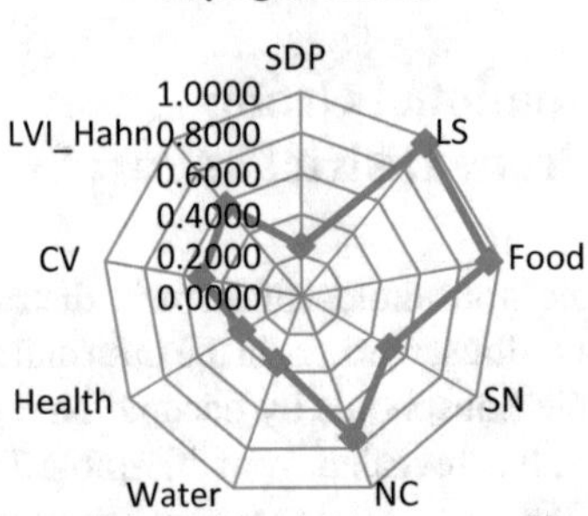

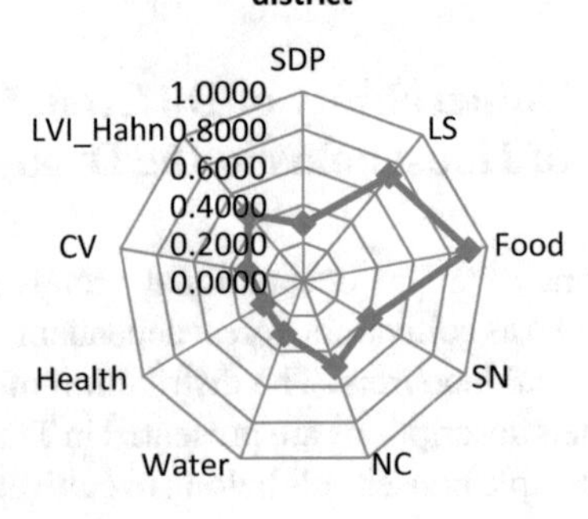

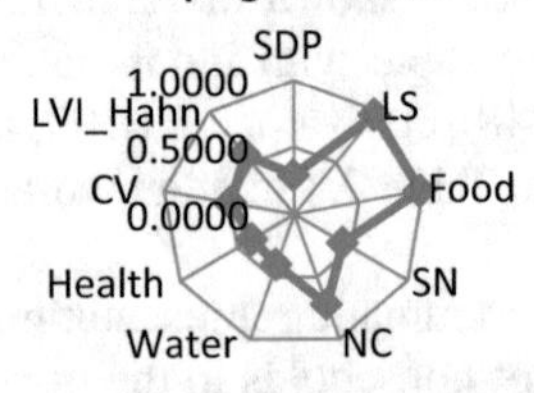

Fig. 7.8 Radar diagram for LVI_Hahn for different livelihood groups in Foothill district of Jalpaiguri

Table 7.6 Indices of major components and vulnerability indices for different livelihood groups in foothill region of Jalpaiguri district, West Bengal

Components of Vulnerability	Cultivators $N_1 = 19$	Forest dependent communities $N_2 = 6$	Workers in the informal sector $N_3 = 75$	Workers in the formal sector $N_4 = 13$	Casual Labourers $N_5 = 17$
Exposure index	0.6455	0.5890	0.5039	0.3041	0.4895
Sensitivity index	0.3932	0.4115	0.4772	0.3461	0.4983
Adaptive capacity index	0.6330	0.6950	0.6669	0.5832	0.6657
LVI_IPCC	**0.5572 (2)**	**0.5652 (1)**	**0.5493 (4)**	**0.4111 (5)**	**0.5512 (3)**

Source Computed by author from primary data; Figures in the parenthesis represent rank

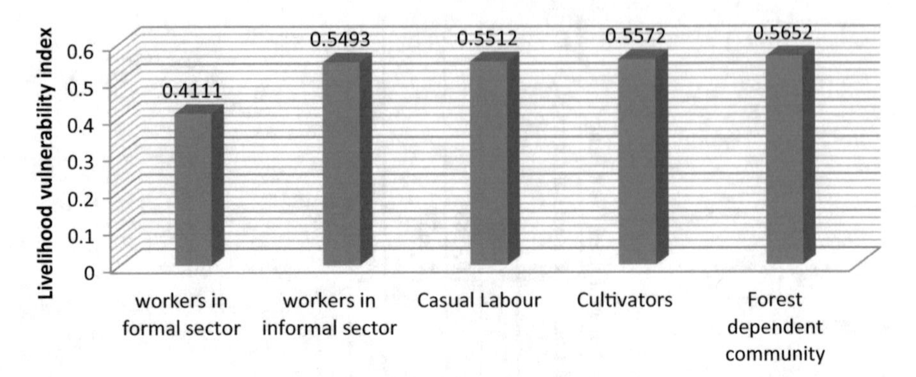

Fig. 7.9 Modified Vulnerability indices (LVI_IPCC) for livelihood groups in Foothill region of Jalpaiguri district

capital index and climate variable index. This means that forest dependent communities more affected by climate and natural capital. The livelihood vulnerability of different sub-components for different occupational group of households is shown in Fig. 7.13.

The modified livelihood vulnerability indices (LVI_IPCC) for different occupational group of households are shown in Table 7.9. It is observed from Table 7.9 that vulnerability of forest dependent communities is found to be highest (0.6243) and lowest for workers in the informal sector (0.5555). The more vulnerable communities are forest dependent communities followed by cultivators, casual labourers and workers in the informal sector in the drought district of Purulia. The causes of vulnerability of forest dependent communities are due to higher values of exposure, higher values of sensitivity and lower value of adaptive capacity. Figure 7.14 shows the diagrammatic representation of modified livelihood vulnerability indices for different occupational group of households in the district of Purulia. The Rader diagram for modified livelihood vulnerability indices of different sub-components for different occupational group of households are shown in Fig. 7.15.

7.4 Vulnerability of Different Occupational Group of Households in the Coastal Region of Sunderban

The sample households of the coastal region of Indian Sundarban are classified by major occupations groups into fishing households, crab collecting households, casual labourers, petty businessmen and workers in the informal sectors. The distribution of sample households by major occupations and their descriptions are presented in Table 7.10. It is revealed that 14% of sample households belong to fishing, 27% of households to crab collection, and 31% belong to informal sectors, 9% to petty businessmen and 19% are casual labourers (see Table 7.10 and Fig. 7.16).

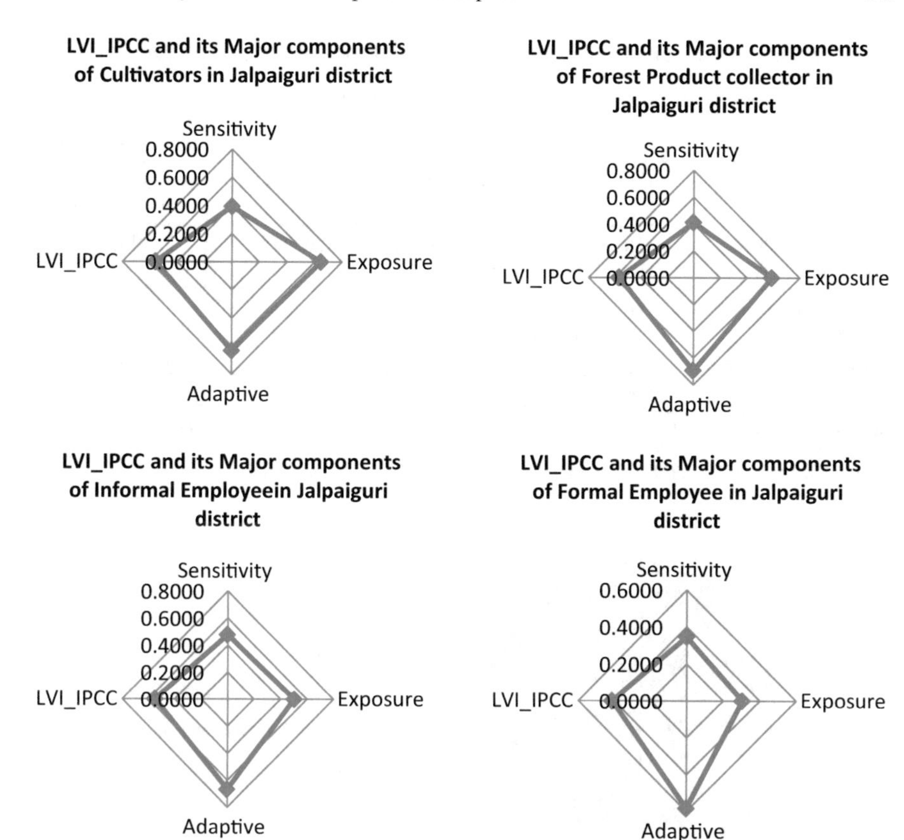

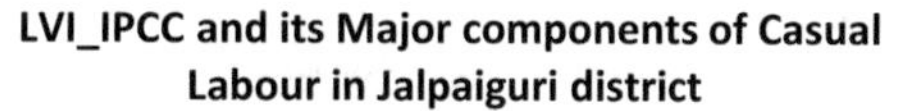

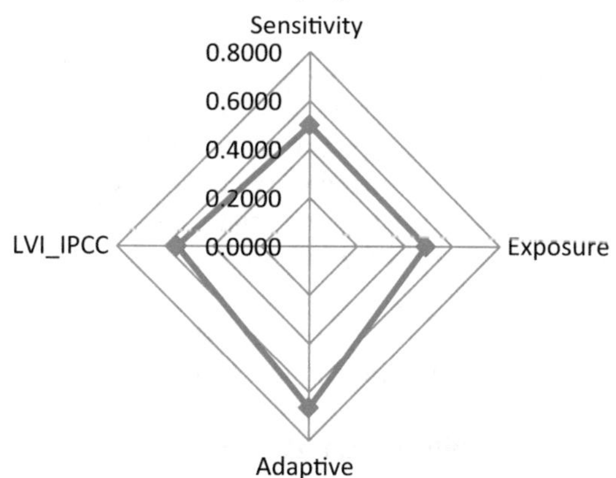

Fig. 7.10 Rader diagram for Modified Vulnerability indices (LVI_IPCC) for different for livelihood groups in Foothill district, Jalpaiguri

Table 7.7 Distribution of different occupational group of households in the Drought prone region of Purulia district

Occupational groups	Description	Sample households (%)
Cultivators (N_1)	Households who are engaged in cultivation of agricultural crops	49 (33)
Forest dependent communities (N_2)	Households who are dependent on forests for livelihood generation	39 (26)
Workers in the informal sector (N_3)	Households who are working in hotel, restaurant, shop and factory etc.	42 (28)
Casual labour (N_5)	Households who are engaged in construction of road, building, drainage etc.	20 (13)
All ($N_1 + N_2 + N_3 + N_4 + N_5$)		**150 (100)**

Source Computed by author from field survey; Figures in the parenthesis represent percentage

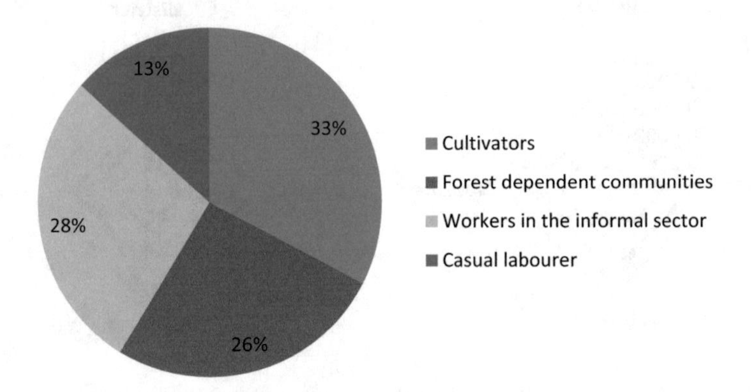

Fig. 7.11 Composition of occupational group of households in Purulia district

The livelihood vulnerability indices for different livelihood groups in coastal Sunderban, West Bengal are presented in Table 7.11. The livelihood vulnerability index of different occupation groups varies from 0.3719 to 0.5431. It is observed from Table 7.11 that the crab collecting communities in coastal Sunderbans are most vulnerable (0.5431) followed by the fishing communities (0.4998), the casual labourer (0.3968), the workers in the informal sectors (0.3719). It is also revealed that the petty businessmen is found to be the least vulnerable (0.3276) community in coastal Sunderbans (Table 7.11). Figure 7.17 shows the diagrammatic representation of vulnerability indices of different occupation group of households in coastal Sunderbans.

The nature of vulnerabilities can be described in terms of different sub-component of vulnerability. The different sub-components of vulnerability are socio-demographic profile (SDP), Livelihood strategy (LS), Food, Social Network, Natural capital, Water, Health and Climate. These sub-components are taken into consideration for measuring livelihood vulnerability index of different occupation groups. The

Table 7.8 Indices of sub components and vulnerability indices for different livelihood groups in Purulia district, West Bengal

Sub-components of vulnerability	Cultivators $N_1 = 49$	Forest dependent community $N_2 = 39$	Workers in informal sector $N_3 = 42$	Casual labour $N_4 = 20$
Socio-demographic profile index	0.3262	0.3508	0.2657	0.4798
Livelihood strategy index	0.6550	0.6369	0.7038	0.7525
Food index	0.9848	0.9808	0.9817	0.9250
Social network index	0.4463	0.4101	0.4677	0.5516
Natural capital index	0.8354	0.9108	0.8498	0.9050
Water index	0.5464	0.6921	0.5971	0.6238
Health index	0.5165	0.4092	0.3083	0.4188
Climate variable index	0.5769	0.6076	0.4767	0.4841
Livelihood vulnerability index (by Hahn et al. 2009)	**0.6109 (3)**	**0.6248 (2)**	**0.5814 (4)**	**0.6426 (1)**

Source Computed by author from primary data; Figures in the parenthesis represent rank

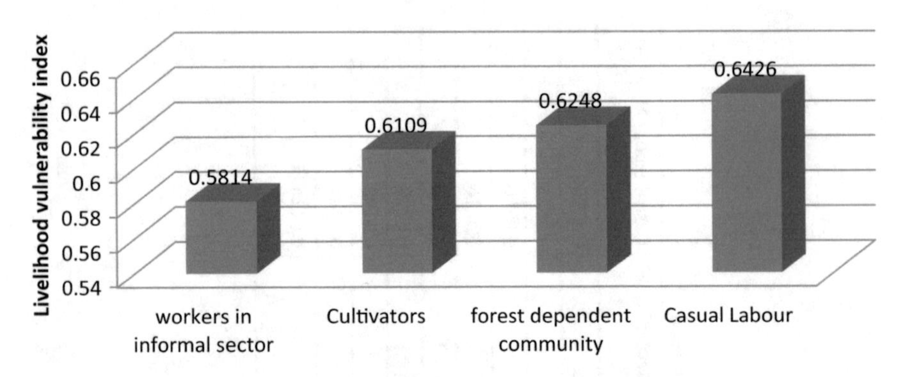

Fig. 7.12 LVI of different livelihood groups in Purulia district, West Bengal

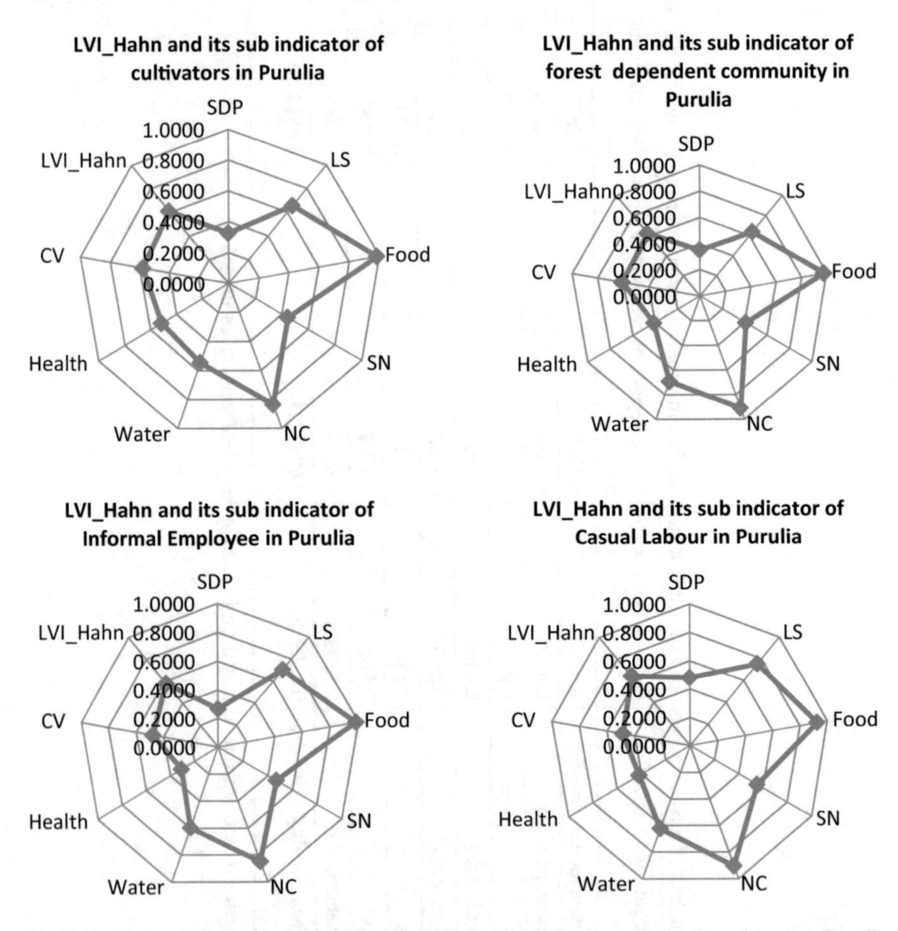

Fig. 7.13 Radar diagram for LVI_Hahn for different livelihood groups in drought prone Purulia district

Table 7.9 Indices of major components and vulnerability indices for different livelihood groups in Purulia district, West Bengal

Components of vulnerability	Cultivators N_1 = 49	Forest dependent community N_2 = 39	Workers in the informal sector N_3 = 42	Casual labour N_4 = 20
Exposure	0.5769	0.6076	0.4767	0.4841
Sensitivity	0.6328	0.6707	0.5851	0.6492
Adaptive	0.6031	0.5947	0.6047	0.6772
LVI_IPCC	**0.6042 (2)**	**0.6243 (1)**	**0.5555 (4)**	**0.6035 (3)**

Source Computed by author from primary data; Figures in the parenthesis represent rank

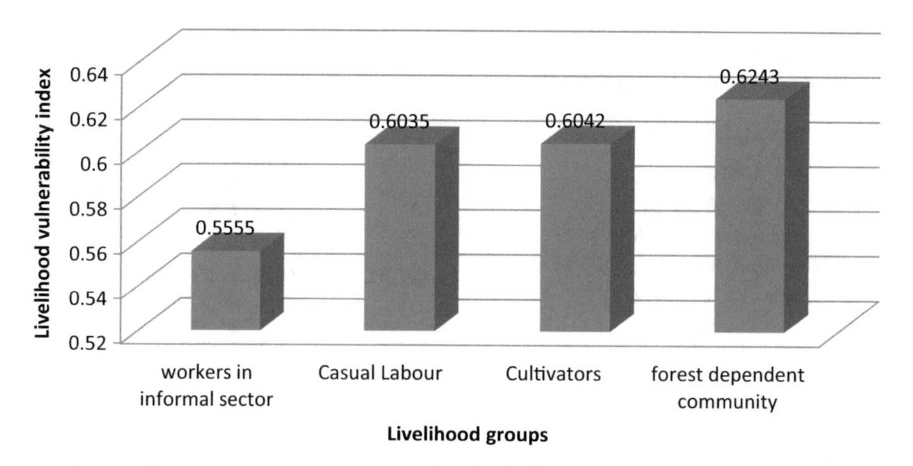

Fig. 7.14 ModifiedVulnerability indices (LVI_IPCC) for livelihood group of households in Purulia district

higher values of these indices mean higher vulnerability and vice versa. The fishing and crab collecting communities are more vulnerable than other occupation groups due to higher values of natural capital, social network, water, health and climate (Table 7.11). The high dependency of natural capital, less involvement in social network, salinity of water, inaccessibility to health care facility and less perception about climatic events are responsible for higher vulnerability of the fishing and crab collecting communities in coastal Sunderbans. The workers in the informal sector and petty businessmen are less affected by natural capital, social network and water (Table 7.11). The livelihood vulnerability of different sub-components for different occupation groups are shown in Fig. 7.18.

The modified livelihood vulnerability indices (LVI_IPCC) for different occupation groups in Sundarbans are shown in Table 7.12. It is observed from Table 7.12 that vulnerability of the crab collecting communities is highest (0.5864) and lowest for petty business men (0.3616). The more vulnerable communities are crab collecting communities followed by fishing communities, casual labour and worker in the

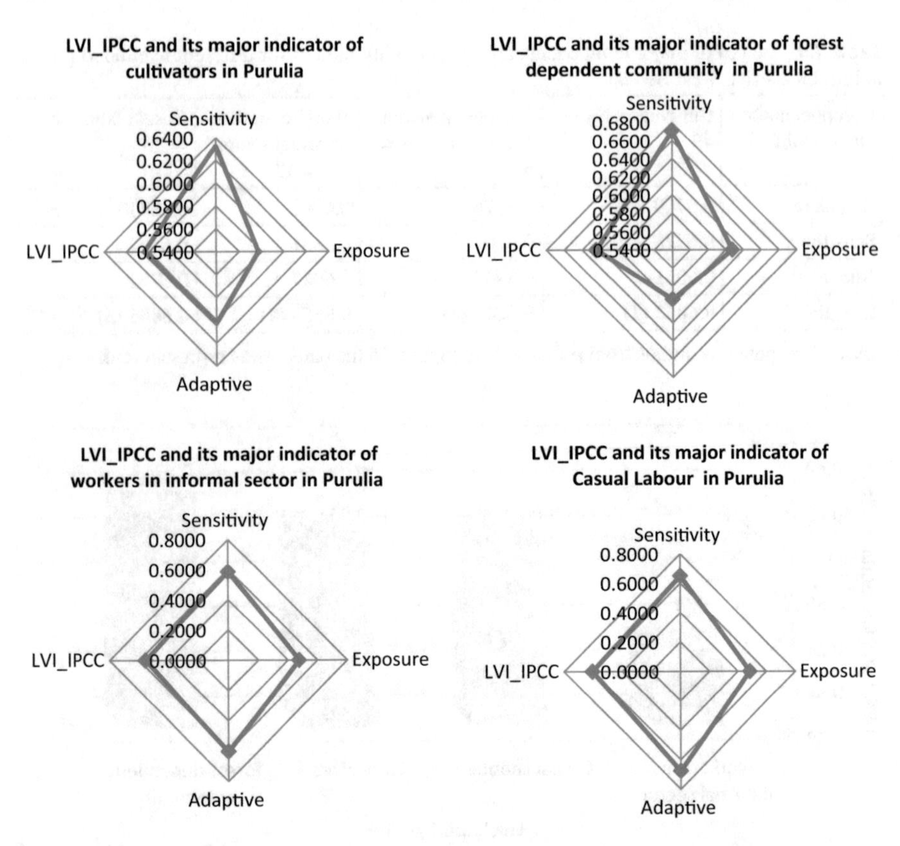

Fig. 7.15 Rader diagram for Modified Vulnerability indices (LVI_IPCC) for different for livelihood group of households in Purulia district

informal sector in the coastal Sunderbans. The fishing and crab collecting communities are facing with high exposure and sensitivity compared to the other occupation groups (Table 7.12). This Table 7.12 shows that the adaptive capacity of each occupation group is lower than exposure and sensitivity components of vulnerability. Figure 7.19 shows the diagrammatic representation of modified livelihood vulnerability indices for different occupation groups in coastal Sunderbans. The modified livelihood vulnerability indices of different sub-components for different occupation groups are shown in Fig. 7.20.

Table 7.10 Distribution of sample for different occupation group of households in the coastal region of Sunderban

	Description	Number of households (%)
Fishing households (N_1)	Households who are engaged in fishing in nearby river	28 (14.22)
Crab collecting households (N_2)	Households who are engaged in collecting crab in nearby river	52 (26.40)
Workers in the informal sector (N_3)	Households who are engaged in rice mill/cold storage/shop/shopping mall in nearby district	61 (30.97)
Petty businessmen (N_4)	Households who are engaged in doing small business like fish arat, crab arat, vending vegetable, small grocery/cycle and bike maintenance etc.	18 (9.14)
Casual labourers (N_5)	Households who are engaged in agricultural work in other's land, construction of road, building etc.	38 (19.29)
All ($N_1 + N_2 + N_3 + N_4 + N_5$)		**197 (100)**

Source Computed by author from field survey; Figures in the parenthesis represent percentage

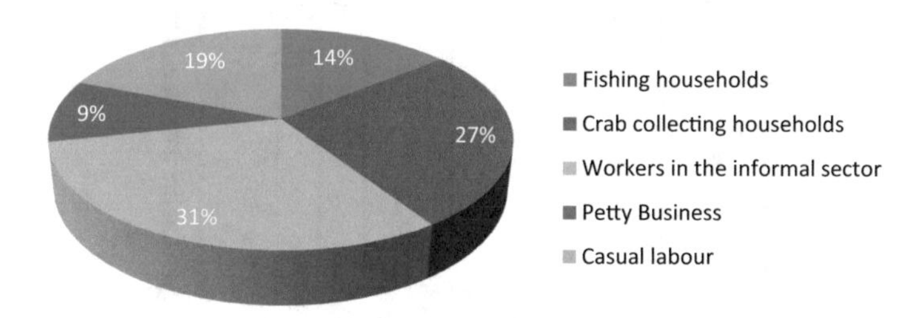

Fig. 7.16 Composition of occupational group of households in Sundarban

7.5 Vulnerability of Different Occupational Group of Households in the Coastal Region of East Midnapore

The sample households of east Midnapore are classified by major occupations into casual labourers, fishing communities, workers in informal sector, workers in the formal sector and van pullers. The distribution of sample households by major occupations and their descriptions is presented in Table 7.13. It is revealed that 18% sample households belong to casual labour, 13% to cultivators, 16% to fishing communities, 13% belongs to informal sectors, 16% to workers in the formal sector, and 24% belongs to van puller (see Table 7.13 and Fig. 7.21).

The livelihood vulnerability indices of different occupational group of households in the coastal region of East Midnapore district are shown in Table 7.14. It is found from this table that casual labourer are most vulnerable in the coastal region

Table 7.11 Indices of sub components and vulnerability indices for different livelihood groups in coastal Sunderban, West Bengal

Sub-components of vulnerability	Fishing communities	Workers in informal sectors	Petty business men	Crab collecting communities	Casual labour
Socio-demographic profile index	0.4260	0.2700	0.2537	0.4725	0.3580
Livelihood strategy index	0.2367	0.3598	0.3585	0.0027	0.1894
Food index	0.02	0.0984	0.2778	0.01	0.1579
Social network index	0.6108	0.5693	0.3497	0.9376	0.5448
Natural capital index	0.8042	0.6486	0.5327	0.9990	0.7239
Water index	0.7781	0.0007	0.0082	0.8151	0.0092
Health index	0.6087	0.5192	0.3682	0.4743	0.5630
Climate variable index	0.5340	0.5089	0.4718	0.6429	0.6279
Livelihood vulnerability index (by Hahn et al. 2009)	**0.4998 (2)**	**0.3719 (4)**	**0.3276 (5)**	**0.5431 (1)**	**0.3968 (3)**

Source Computed by author from field survey primary data; Figures in the parenthesis represent rank

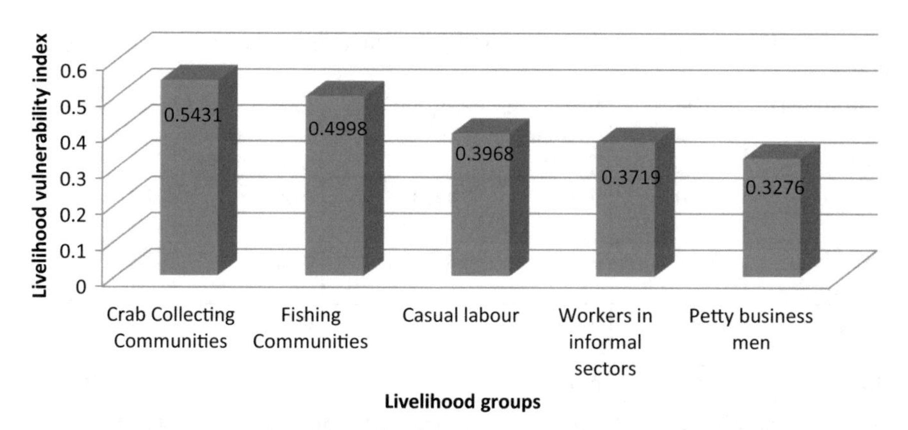

Fig. 7.17 LVI of different occupational groups in coastal Sunderban, West Bengal

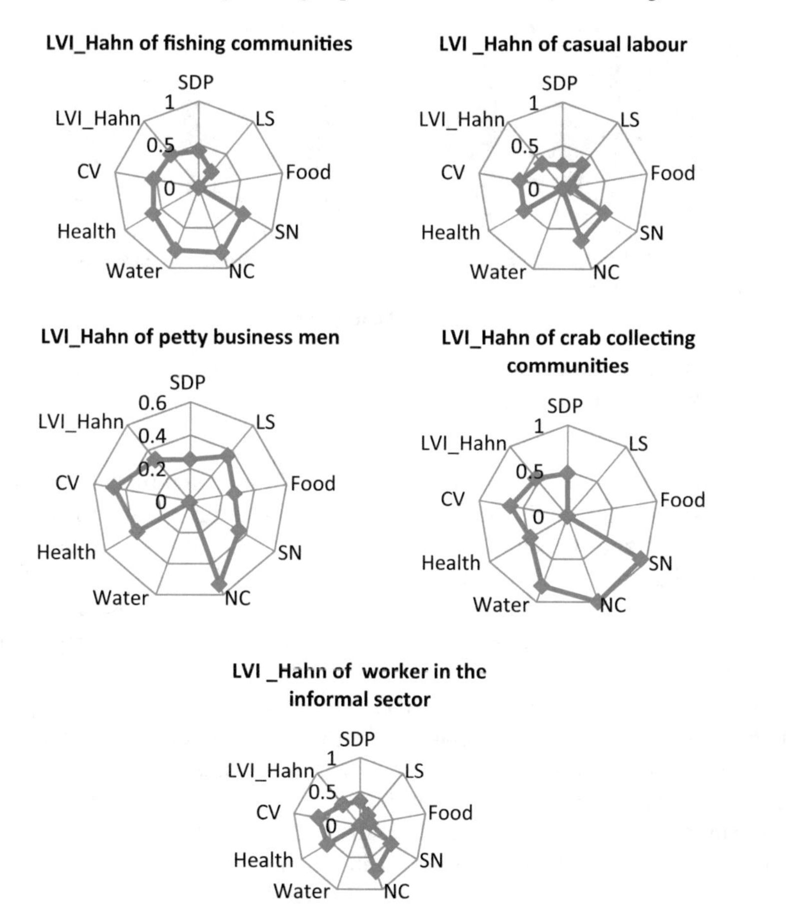

Fig. 7.18 Rader diagram for LVI_Hahn for different occupation groups in coastal Sunderbans, West Bengal

Table 7.12 Modified Vulnerability indices (LVI_IPCC) for different livelihood groups in coastal Sunderbans

Components of vulnerability	Fishing community	Workers in informal sectors	Petty business men	Crab collecting communities	Casual labour
Exposure index	0.5340	0.5089	0.4718	0.6429	0.6279
Sensitivity index	0.7303	0.3895	0.3030	0.7631	0.4321
Adaptive index	0.3184	0.3244	0.3099	0.3532	0.3125
LVI_IPCC	**0.5276 (2)**	**0.4076 (4)**	**0.3616 (5)**	**0.5864 (1)**	**0.4575 (3)**

Source Computed by author from field survey; Figures in the parenthesis represent rank

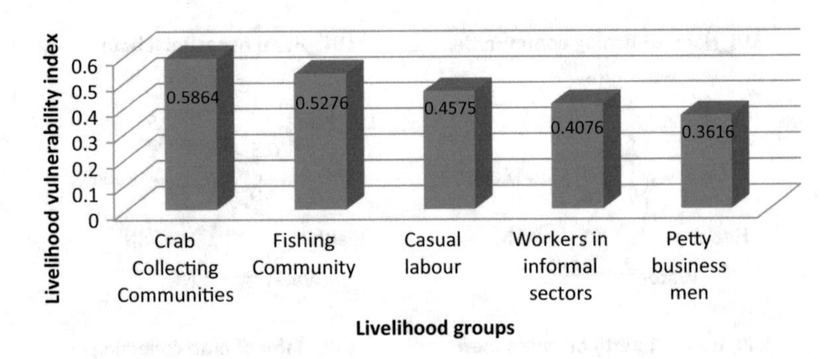

Fig. 7.19 Modified Vulnerability indices (LVI_IPCC) for occupational groups in coastal Sunderban, West Bengal

of East Midnapore district with index value 0.5878 followed by fishing communities (0.5741),Van Puller (0.5070), workers in the informal sectors (0.4872), cultivators (0.4785) and workers in the formal sectors (0.4756). The least vulnerable occupational group is workers in the formal sector in the East Midnapore district (see Fig. 7.22). Figure 7.22 shows the diagrammatic representation of vulnerability indices of different occupational group of households in the coastal region of East Midnapore district.

The nature of vulnerabilities is described in terms of different sub-component of vulnerability like socio-demographic profile (SDP), Livelihood strategy (LS), Food, Social Network, Natural capital, Water, Health and Climate. The higher values of these indices mean higher vulnerability and vice versa. The casual labourers are more vulnerable than other occupational groups due to higher values of socio-demographic profile index, food index, social network index, natural capital index and of health index (Table 7.14). The high dependency of natural capital, inaccessibility to food, poor socio-demographic conditions, and inaccessibility of health facilities are responsible for higher vulnerability. The second most vulnerable community is the fishing

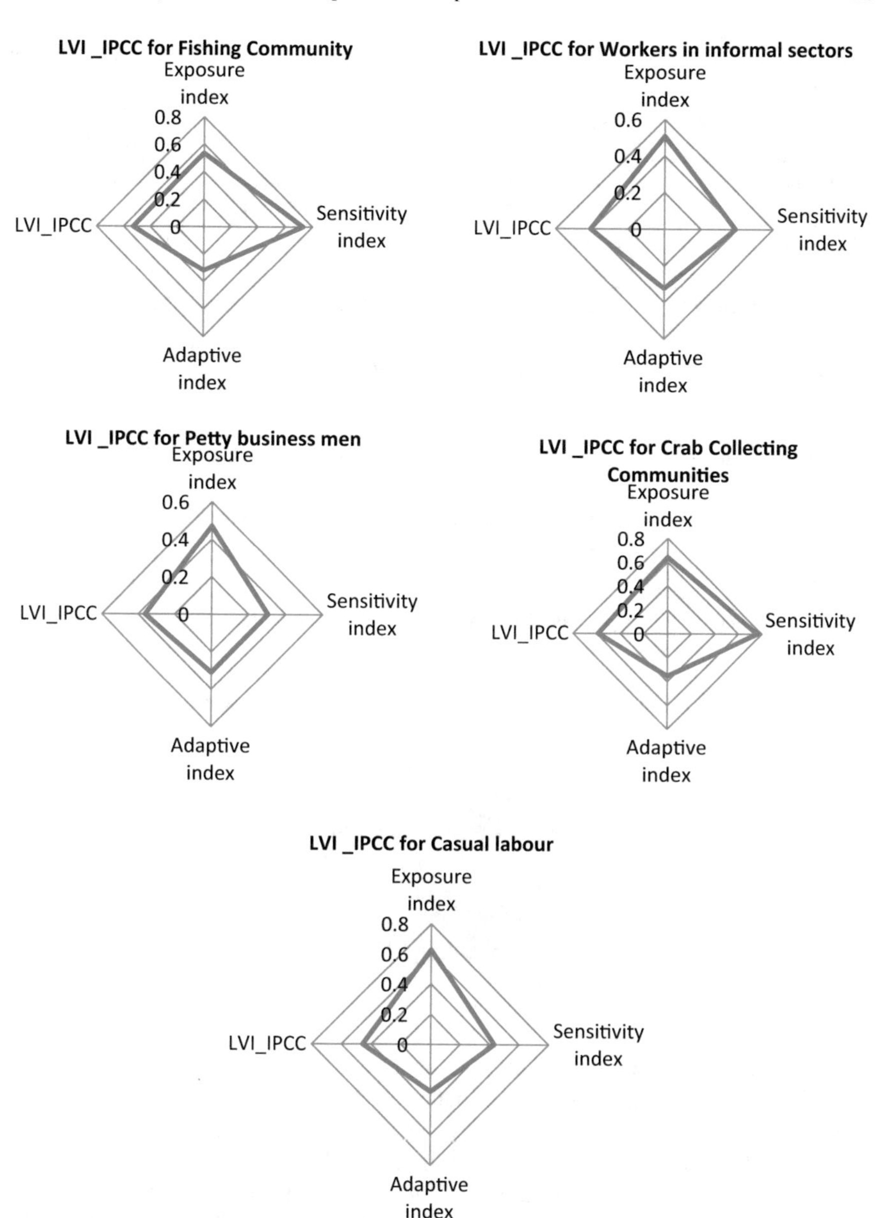

Fig. 7.20 Rader diagram for Modified Vulnerability indices (LVI_IPCC) for different occupation groups in coastal Sunderbans, West Bengal

Table 7.13 Distribution of different occupational group of households in the coastal region of East Midnapore district

Occupational groups	Description	Sample households (%)
Casual labour (N_1)	Households who are engaged in construction activities like roads, buildings, drainage etc.	28 (18)
Cultivators (N_2)	Households' whose main occupation is cultivation	20 (13)
Fishing community (N_3)	Households who are engaged in fishing in the sea	26 (16)
Workers in informal sector (N_4)	Households who are working in hotels, restaurants, shops, shopping malls and factories etc.	21 (13)
Workers in formal sector (N_5)	Households who are working in government job	26 (16)
Van puller (N_6)	Households who are engaged in motor van for passenger/fishing activities	38 (24)
All ($N_1 + N_2 + N_3 + N_4 + N_5 + N_6$)		**159 (100)**

Source Computed by author from field survey; Figures in the parenthesis represent percentage

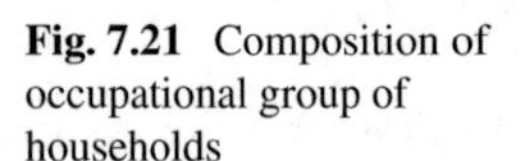

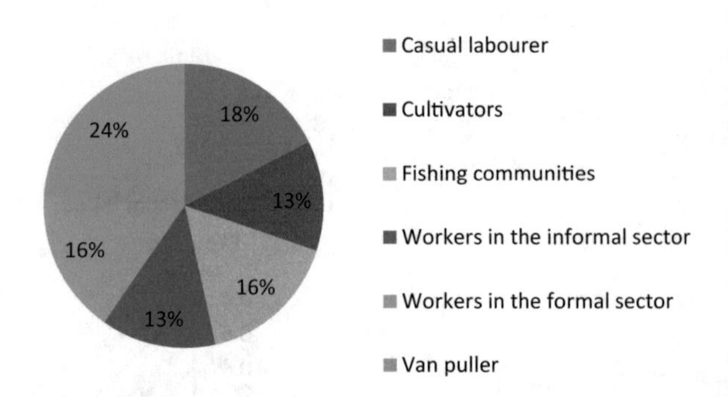

Fig. 7.21 Composition of occupational group of households

communities. They are affected by poor socio-demographic conditions, inaccessibility of food, water and health, and climate variable. The livelihood vulnerability of different sub-components for different occupational groups is shown in Fig. 7.23.

The modified livelihood vulnerability indices (LVI_IPCC) for different occupational group of households are shown in Table 7.15. It is observed from Table 7.15 that vulnerability of the casual labourers is found to be highest (0.5108) and lowest for the workers who are in the informal sectors (0.4002). The more vulnerable communities are casual labourer followed by fishing communities, van puller, workers in the formal sector, cultivators and worker in the informal sectors in the coastal region of East Midnapore. The casual labourer and fishing communities are facing with high sensitivity and low adaptive capacity compared to the other occupational group of households (Table 7.15). Figure 7.24 shows the diagrammatic representation of

Table 7.14 Indices of sub components and vulnerability indices for different livelihood groups in coastal region of East Midnapore, West Bengal

Sub-components of Vulnerability	Casual Labour	Fishing community	Van puller	Workers in informal sectors	Cultivator	Workers in the formal sectors
Socio-demographic profile index	0.8381	0.7201	0.5785	0.4663	0.4999	0.6020
Livelihood strategy index	0.6036	0.5582	0.5467	0.5908	0.4972	0.6667
Food index	0.9750	0.9785	0.9742	0.9902	0.9299	0.7623
Social net work index	0.5901	0.4088	0.5630	0.5625	0.4926	0.3582
Natural capital index	0.8979	0.7586	0.8171	0.8445	0.8089	0.8695
Water index	0.1268	0.2717	0.1178	0.1178	0.2037	0.0501
Health index	0.3476	0.3168	0.1402	0.0873	0.1734	0.1228
Climate variable index	0.3234	0.3641	0.3185	0.1984	0.2227	0.3736
Livelihood vulnerability index (by Hahn et al. 2009)	**0.5878 (1)**	**0.5471 (2)**	**0.5070 (3)**	**0.4822 (4)**	**0.4785 (5)**	**0.4756 (6)**

Source Computed by author from field survey primary data; Figures in the parenthesis represent rank

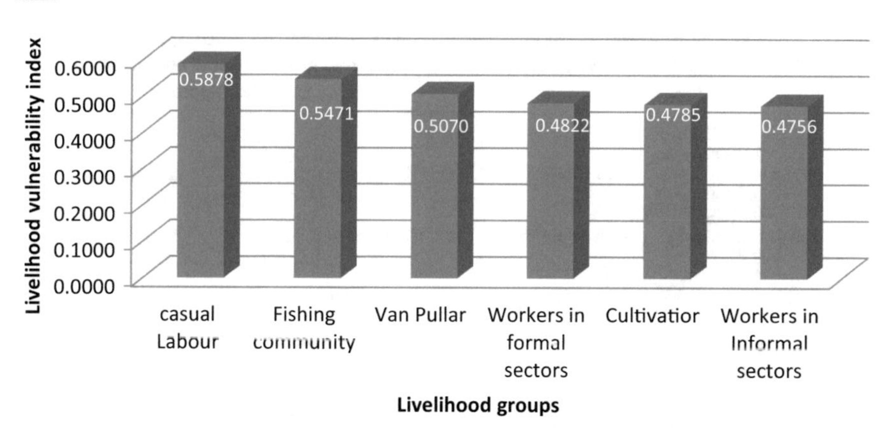

Fig. 7.22 LVI of different livelihood groups in coastal region of East Midnapore, West Bengal

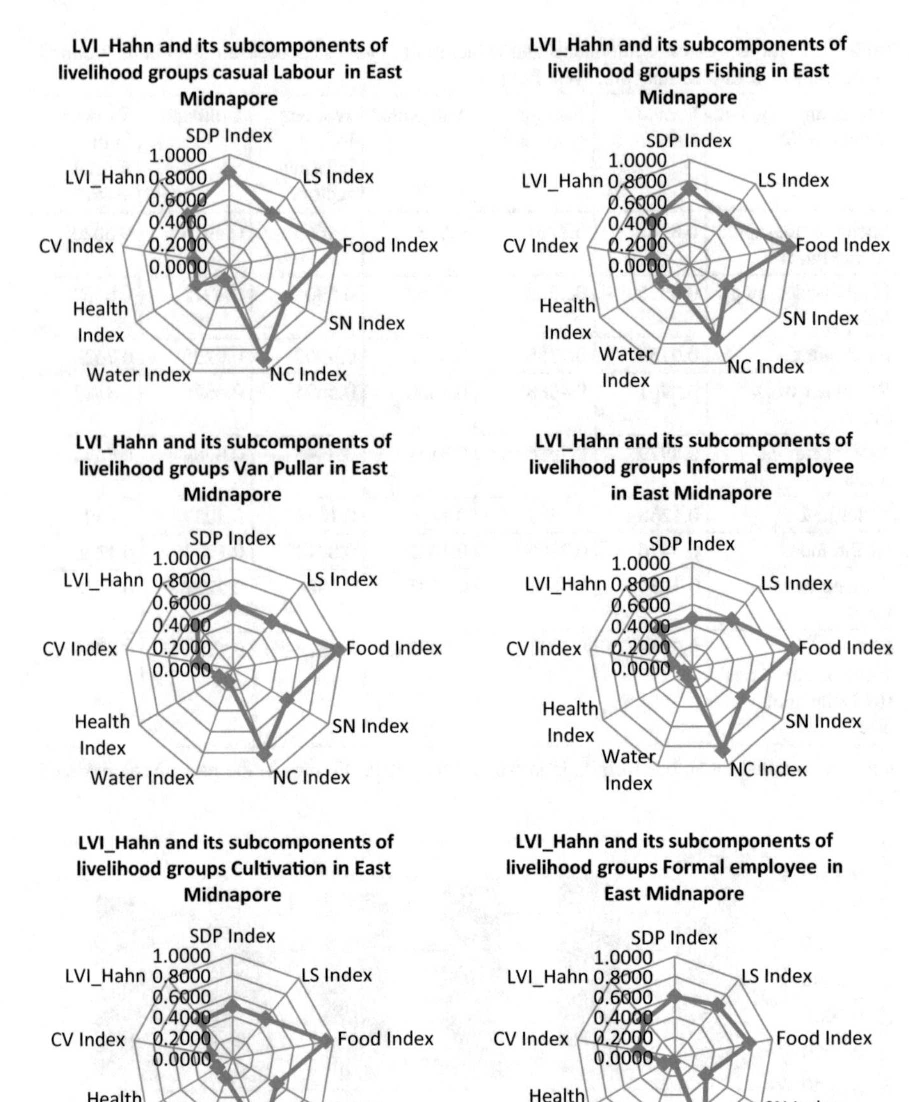

Fig. 7.23 Radar diagram for LVI_Hahn for different livelihood groups in coastal region of East Midnapore, West Bengal

Table 7.15 Modified Vulnerability indices (LVI_IPCC) for different livelihood group in coastal region of East Midnapore

Components of vulnerability	Casual labour	Fishing community	Van puller	Workers in formal sectors	Cultivators	Workers in the informal sectors
Exposure	0.3234	0.3641	0.3185	0.3736	0.2227	0.1984
Sensitivity index	0.4574	0.4490	0.3584	0.3475	0.3953	0.3498
Adaptive index	0.7517	0.6664	0.6656	0.5973	0.6049	0.6524
LVI_IPCC	**0.5108 (1)**	**0.4932 (2)**	**0.4475 (3)**	**0.4395 (4)**	**0.4077 (5)**	**0.4002 (6)**

Source Computed by author from field survey; Figures in the parenthesis represent rank

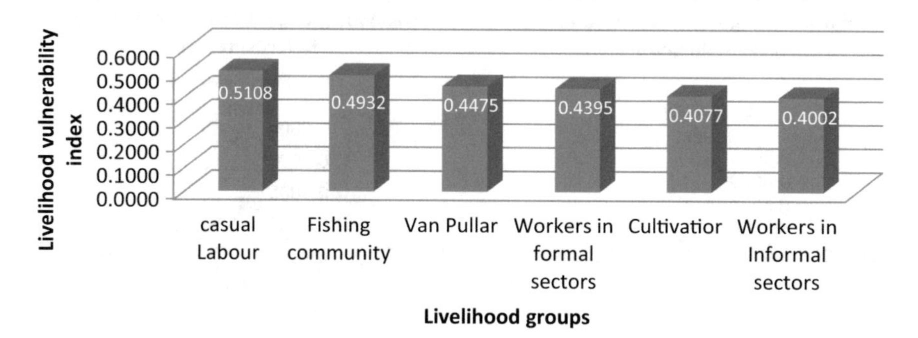

Fig. 7.24 Modified Vulnerability indices (LVI_IPCC) for occupational groups in coastal region of East Midnapore, West Bengal

modified livelihood vulnerability indices for different occupational group of households in coastal region of East Midnapore. Figure 7.24 shows the diagrammatic representation of livelihood vulnerability indices for different occupational group of households in the East Midnapore district. The Rader diagram for modified livelihood vulnerability indices of different sub-components for different occupational group of households are shown in Fig. 7.25.

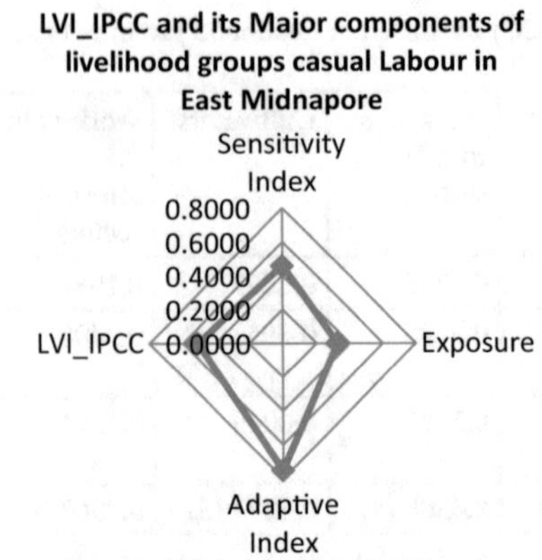

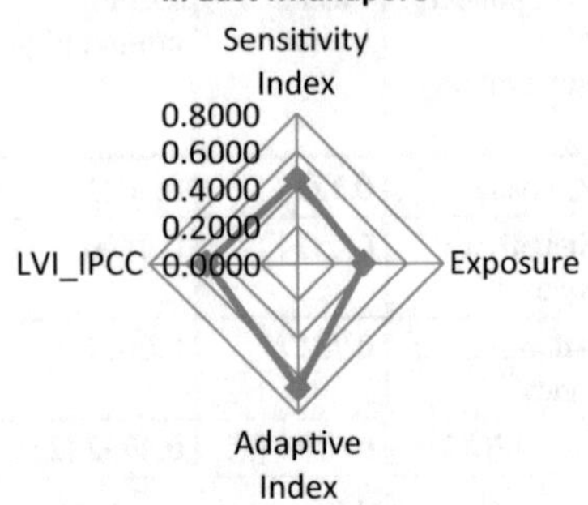

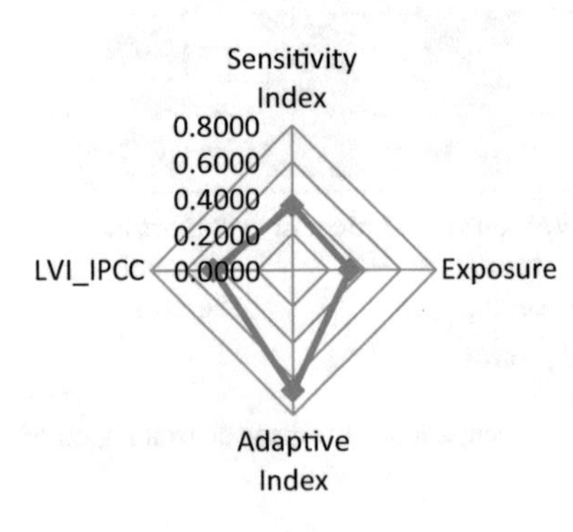

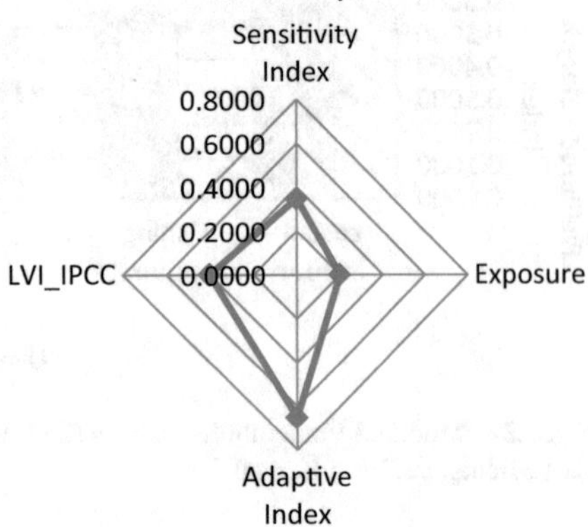

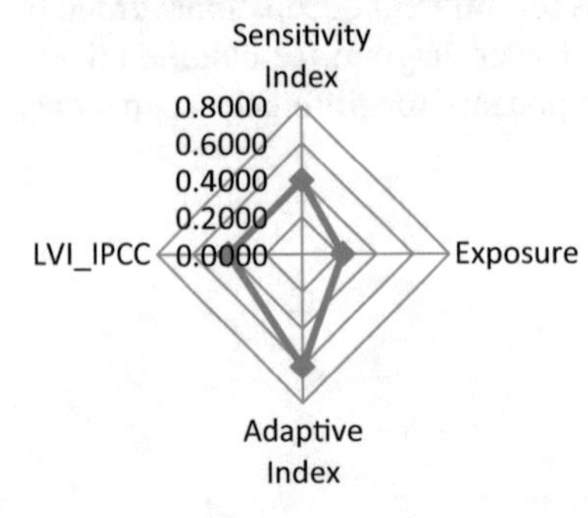

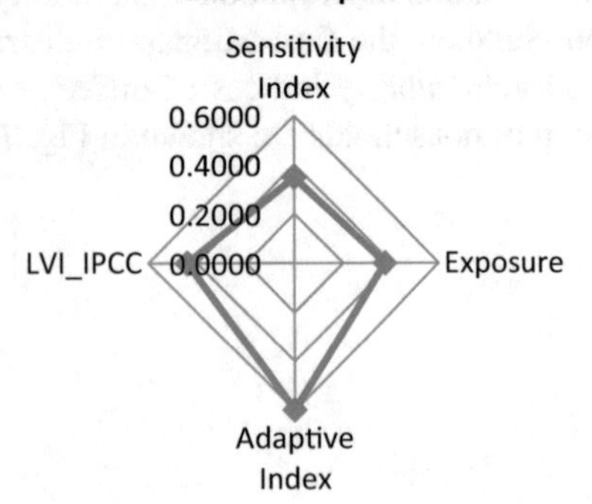

Fig. 7.25 Rader diagram for Modified Vulnerability indices (LVI_IPCC) for different occupation groups coastal region of East Midnapore, West Bengal

References

Hahn MB, Riederer AM, Foster SO (2009) The livelihood vulnerability index: a pragmatic approach to assessing risks from climate variability and change—a case study in Mozambique. Global Environ Change 19(1):74–88. https://doi.org/10.1016/j.gloenvcha.2008.11.002

IPCC (2007) Climate change (2007). Impacts adaptation and vulnerability. Summary for policy makers. Intergovernmental Panel on Climate Change (IPCC) http://www.ipcc.cg/SPM.pdf

Chapter 8
Measurement of Gender-Wise Vulnerability in Different Agro-Climatic Regions of West Bengal

This chapter measures gender-wise vulnerability to climate change at the household level across five agro-climatic regions of West Bengal. The Livelihood Vulnerability Index (LVI) and combined LVI-IPCC index have been used to measure such gender vulnerability.

8.1 Gender-Wise Measurement of Vulnerability in the Hill Region of Darjeeling District

We have used livelihood vulnerability index of Hahn et al. 2009 and modified LVI-IPCC index to measure vulnerability. The indicators used in this calculation are described in sub-Sect. 3.4 of the Chap. 3.

In the hill region of Darjeeling district the total sample comprises150 households out of which 75 households are male headed and 75 households are female headed households.

The livelihood vulnerability indices for female and male headed households are found to be 0.4556 and 0.4483 respectively in the hill region of Darjeeling district shown in Table 8.1. It shows that female headed households are more vulnerable than male headed households in Darjeeling district. The results of two sample t tests reveal that mean differences of livelihood vulnerability indices between female and male headed households are statistically significant at 5% level.

The results of combined livelihood vulnerability indices (LVI_IPCC) for female and male headed households along with t-values in the hill region of Darjeeling district are presented in Table 8.2. The combined livelihood indices for female and male are found to be 0.5109 and 0.5057 respectively in the hill region of Darjeeling district (Table 8.2).

The result shows that females are more vulnerable than male. The female households are more vulnerable than male headed households are due to higher values SDP

J. P. Basu, *Climate Change Vulnerability and Communities in Agro-climatic Regions of West Bengal, India*, https://doi.org/10.1007/978-3-030-50468-7_8

Table 8.1 Sub-components wise Livelihood vulnerability index for gender with t-test in hill region of Darjeeling district

Sub-component	Index for female (N = 75)	Index for male (N = 75)	t-value	P value
SDP index	0.2829	0.2454	1.7483	0.05179
LS index	0.2365	0.2542	−1.0459	0.2973
Food index	0.0145	0.0391	−1.3702	0.1727
SN index	0.6868	0.6438	2.9927	0.01225
NC index	0.7273	0.7261	0.049	0.961
Water index	0.7928	0.7495	1.8459	0.0604
Health index	0.2749	0.2976	−0.5763	0.5653
CV index	0.6293	0.6303	−0.0243	0.9807
Overall livelihood vulnerability index	**0.4556**	**0.4483**	**1.7416**	**0.0595**

Sources Computed by author from field survey data

Table 8.2 LVI_IPCC indices of Gender by contributory factors in the hill region of Darjeeling district

Contributory factor	Computed index for female	Computed index for male	t-value	P value
Sensitivity index	0.5983	0.5911	0.5683	0.5707
Exposure	0.6293	0.6303	−0.0243	0.9807
Adaptive index	0.3052	0.2956	0.6871	0.4931
LVI_IPCC	**0.5109**	**0.5057**	**2.3454**	**0.0303**

Sources Computed by author from field survey data

index, Social Network index and water index. The livelihood vulnerability indices (in terms of both measures) for female and male are shown in Fig. 8.1.

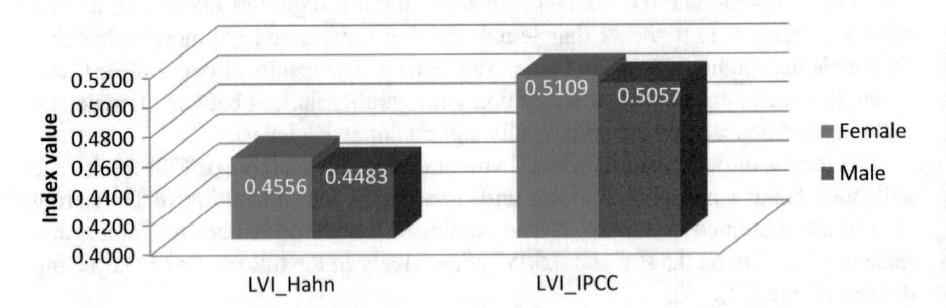

Fig. 8.1 LVI and LVI_IPCC for female and male in the hill region of Darjeeling district

8.2 Gender-Wise Measurement of Vulnerability in Foothill Region of Jalpaiguri District

In the foot-hill region of Jalpaiguri district, the total sample comprises130 households out of which 76 households are male and 54 households are female headed households.

The livelihood vulnerability indices for female and male-headed households are found to be 0.4090 and 0.4175 respectively in the foot-hill region of Jalpaiguri district. It shows that male headed households are more vulnerable than female headed households in the foothill region of Jalpaiguri district. On the basis of two sample t tests, the mean differences of livelihood vulnerability indices between female and male headed households are statistically significant. The male households are more vulnerable than female households are due to higher values livelihood strategy index, Social Network index and climate index (Table 8.3).

The results of combined livelihood vulnerability indices (LVI_IPCC) for female and male headed households along with t-values in the foothill region of Jalpaiguri district are presented in Table 8.4. It is found from Table 8.4 that the combined livelihood vulnerability indices (LVI_IPCC) for female and male headed households are 0.4384 and 0.4701 respectively. This shows that the male headed households are more vulnerable than female headed households and this result is significant. The high vulnerability of male headed households is explained by high exposure index and high adaptive capacity index compared to female headed households.

The livelihood vulnerability indices (in terms of both measures) for female and male are shown in Fig. 8.2.

Table 8.3 Sub-components wise livelihood vulnerability Index for gender with t-tests in the foothill region of Jalpaiguri district

Sub-component	Index for female N = 54	Index for male N = 76	t-value	P value
SDP index	0.2576	0.2138	1.9536	0.0529
LS index	0.1346	0.1755	−2.3709	0.0192
Food index	0.0062	0.0428	−1.7697	0.0792
SN index	0.4309	0.5061	−1.4681	0.1445
NC index	0.8195	0.7702	1.6238	0.1069
Water index	0.8645	0.8199	1.4221	0.1574
Health index	0.3181	0.2491	1.4311	0.1548
CV index	0.4404	0.5628	−3.3388	0.0011
Overall livelihood vulnerability index	**0.4090**	**0.4175**	**−2.5994**	**0.0499**

Sources Computed by author from field survey data

Table 8.4 LVI_IPCC indices of gender by contributory factors in the foot-hill region of Jalpaiguri district

Contributory factor	Computed index for female	Computed index for male	t-value	P value
Sensitivity index	0.6674	0.6130	2.1353	0.0346
Exposure	0.4404	0.5628	−3.3388	0.0011
Adaptive index	0.2073	0.2345	−1.7048	0.0906
LVI_IPCC	**0.4384**	**0.4701**	**−1.9562**	**0.0526**

Sources Computed by author from field survey data

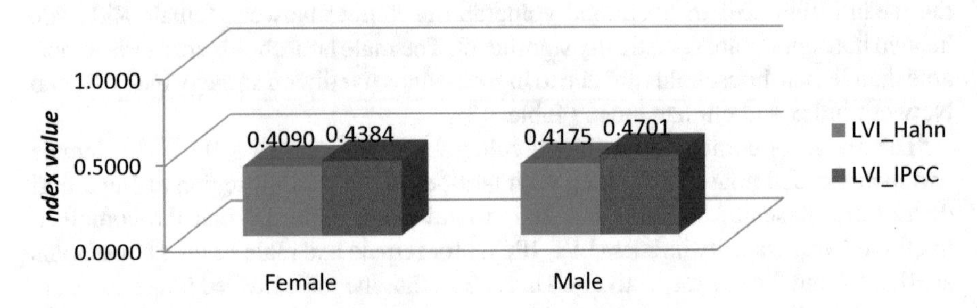

Fig. 8.2 LVI Hahn and LVI IPCC between male and female in foot-hill region of Jalpaiguri district

8.3 Gender-Wise Measurement of Vulnerability in the Drought Region of Purulia District

In the drought region of Purulia district, the total sample comprises150 households out of which 117 households are male headed and 33 households are female headed households.

The results of livelihood vulnerability indices for female and male-headed households in the drought prone district are presented in Table 8.5. It is revealed from Table 8.5 that the vulnerability indices for female and male-headed households are found to be 0.5909 and 0.5237 respectively. This means that female households are more vulnerable than male headed households in the drought region of Purulia district. On the basis of t-values the result is significant. This means that the null hypothesis of no difference in vulnerability indices for female and male headed households is rejected. This further implies that the female households are more vulnerable than male headed households. It is also revealed from Table 8.5 that on the basis of t-values there is a significant difference results observed in computed indices of socio-demographic profile, livelihood strategy, food, social networks, natural capital, water and health except climate component for female-headed and male-headed households. The causes of such high vulnerability in the female headed households are due to the persistence of illiteracy and involvement

Table 8.5 Sub-components wise Livelihood vulnerability Index for gender with t-tests in the drought region of Purulia district

Sub-components	Index		t-test	
	Female	Male	t-value	P value
SDP index	0.536	0.2816	9.0305	0.0000
LS index	0.2641	0.2449	2.17	0.03357
Food index	0.4545	0.3248	1.9768	0.0707
SN index	0.5676	0.5078	2.3001	0.0956
NC index	0.9245	0.9079	1.79	0.08900
Water index	0.7606	0.6532	2.4688	0.0100
Health index	0.4691	0.4479	7.00	0.000
Climate index	0.5992	0.5618	0.7459	0.4600
Overall livelihood vulnerability index	**0.5909**	**0.5237**	**2.9324**	**0.0039**

Sources Computed by author from field survey data

of female earning members in the female households than the male headed households, female households do not have communication skill and few female headed households are associated with the formation of SHGs.

The results of combined livelihood vulnerability indices (LVI_IPCC) for female and male headed households along with t-values of the district of Purulia are presented in Table 8.6. The results showed that the overall vulnerability indices for female and male headed households are found to be 0.572 and 0.4912 respectively (Table 8.6). On the basis of these values the result showed that the female households are more vulnerable than male headed households and their mean difference is also significant. The results of two sample t-tests indicate that there exists a significant difference in female and male headed households in terms of sensitivity and adaptive capacity except exposure components of vulnerability (Table 8.6). The comparative assessment of the vulnerability indices (LVI and IPCC) for female-headed and male-headed households is presented in Fig. 8.3. From Fig. 8.3 it is found that the female households are more vulnerable than male headed households based on both measures of vulnerability assessment.

Table 8.6 LVI_IPCC indices of gender by contributory factors in the drought prone region of Purulia district

Contributory factor	Computed index for female	Computed index for male	t-values	P values
Exposure	0.5992	0.5618	0.7459	0.4569
Sensitivity	0.7181	0.6697	1.9257	0.0561
Adaptive capacity	0.4556	0.3397	4.8132	0.000
LVI_IPCC	**0.572**	**0.4912**	**4.3212**	**0.000**

Sources Computed by author from field survey data

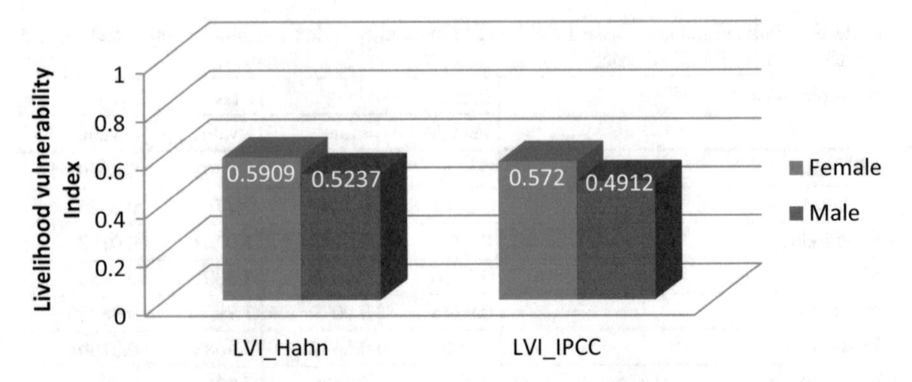

Fig. 8.3 LVI and LVI_IPCC for female and male in the district of Purulia

8.4 Gender-Wise Measurement of Vulnerability in Coastal Region of Sundarban

In the coastal region of Sunderban, the total sample comprises197 households out of which 132 households are male headed and 65 households are female headed households.

The livelihood vulnerability indices for female and male-headed households are found to be 0.5600 and 0.4246 respectively in the coastal region of Sunderbans (Table 8.7). It shows that female headed households are more vulnerable than male headed households in the coastal region of Sunderban. On the basis of two sample t tests, the mean differences of livelihood vulnerability indices between female and male headed households are statistically significant. The higher value of vulnerability

Table 8.7 Sub-components wise Livelihood vulnerability Index for gender with t-test in the coastal region of Sundarban

Sub-component	Index for female N = 65	Index for male N = 132	t-values	P values
SDP index	0.9981	0.7969	6.3245	0.0000
LS index	0.2504	0.2816	−2.1824	0.0303
Food index	0.0769	0.0909	−0.3271	0.7439
SN index	0.6668	0.6120	1.0644	0.2885
NC index	0.7692	0.7661	0.064	0.949
Water index	0.6040	0.5654	1.3522	0.1779
Health index	0.5760	0.4874	2.0287	0.0438
CV index	0.5384	0.5761	−0.9262	0.3555
Overall livelihood vulnerability index	**0.5600**	**0.4226**	**9.9269**	**0.0000**

Sources: Computed by author from field survey primary data

Table 8.8 LVI_IPCC indices of gender by contributory factors in the coastal region of Sundarban

Contributory factor	Computed index for female	Computed index for male	t-values	P values
Sensitivity index	0.6498	0.6063	1.6868	0.0932
Exposure	0.5384	0.5761	−0.9262	0.3555
Adaptive index	0.4981	0.2464	17.1294	0.0000
LVI_IPCC	**0.5621**	**0.4763**	**4.7788**	**0.0000**

Sources Computed by author from field survey data

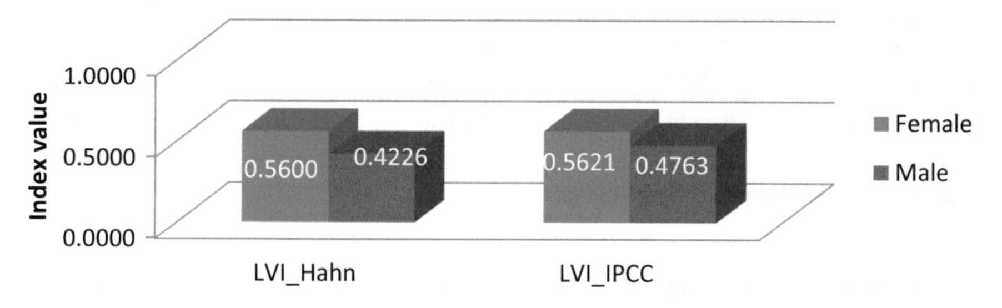

Fig. 8.4 LVI Hahn and LVI IPCC between male and female in coastal Sundarban

for female headed households is due to higher values of socio-demographic profile index, social networks index, natural capital index, water index and health index.

The results of combined livelihood vulnerability indices (LVI_IPCC) for female and male headed households along with t-values of the coastal region of Sunderban are presented in Table 8.8. The results showed that the combined vulnerability indices for female and male headed households are found to be 0.5621 and 0.4763 respectively (Table 8.8). This means that the female headed households are more vulnerable than male headed households and their mean difference is also significant. The results of two sample t-tests indicate that there exists a significant difference in female and male headed households in terms of sensitivity and adaptive capacity except exposure components of vulnerability (Table 8.8). The comparative assessment of the vulnerability indices (LVI and IPCC) for female-headed and male-headed households is presented in Fig. 8.4.

8.5 Gender-Wise Measurement of Vulnerability in Coastal Region of East Midnapore

In the coastal region of East Midnapore, the total sample comprises159 households out of which 135 households are male headed and 24 households are female headed households.

Table 8.9 Sub-components wise Livelihood vulnerability Index for gender with t-test in the coastal region of East Midnapore district

Sub-component	Index for female N = 24	Index for male N = 135	t-value	P value
SDP index	0.3280	0.0979	10.768	0.000
LS index	0.5158	0.5263	−0.3777	0.6469
Food index	0.9583	0.8370	1.5587	0.0605
SN index	0.5938	0.6294	−0.6621	0.7455
NC index	0.9461	0.8449	2.6008	0.0051
Water index	0.2024	0.1334	1.7763	0.0776
Health index	0.2079	0.1336	1.4483	0.0748
CV index	0.5840	0.5361	0.9178	0.1801
Overall vulnerability index	**0.5420**	**0.4673**	**4.1252**	**0.000**

Sources Computed by author from field survey data

The livelihood vulnerability indices for female and male headed households are found to be 0.5420 and 0.463 respectively in the coastal region of East Midnapore shown in Table 8.9. It shows that female headed households are more vulnerable than male headed households in the East Midnapore district. The results of two sample t tests show that mean differences of livelihood vulnerability indices between female and male headed households are statistically significant. The high vulnerability of the female headed households is explained by the higher values of socio-demographic profile index, of food index, of natural capital index, of water index, health index and climate variable index of the sub-components of vulnerability.

The results of combined livelihood vulnerability indices (LVI_IPCC) for female and male headed households along with t-values of the district of East Midnapore are presented in Table 8.10. The results showed that the overall vulnerability indices for female and male headed households are found to be 0.5450 and 0.4673 respectively (Table 8.10). On the basis of these values the result showed that the female households are more vulnerable than male headed households and their mean difference is also

Table 8.10 LVI_IPCC indices of gender by contributory factors in the coastal region of East Midnapore district

Contributory factor	Computed index for female	Computed index for male	t-value	P value
Sensitivity index	0.4521	0.3706	3.2096	0.0008
Exposure	0.5840	0.5361	0.9178	0.1801
Adaptive index	0.5990	0.5227	2.7359	0.0035
LVI_IPCC	**0.5450**	**0.4765**	**3.3456**	**0.0005**

Sources Computed by author from field survey data

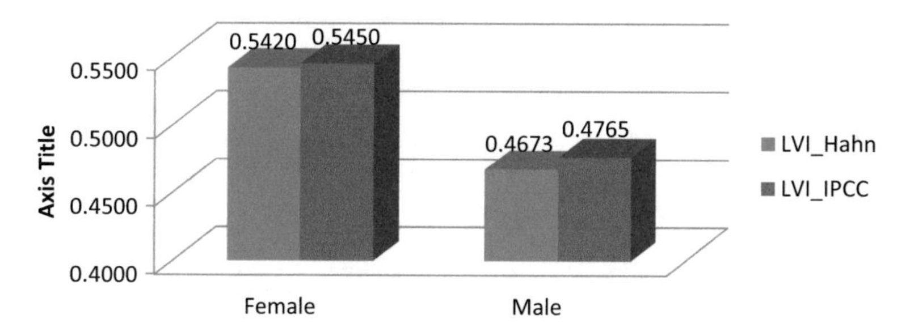

Fig. 8.5 LVI Hahn and LVI-IPCC between male and female in coastal East Midnapore

significant. The results of two sample t-tests indicate that there exists a significant difference in female and male headed households in terms of sensitivity and adaptive capacity except exposure components of vulnerability (Table 8.10). The comparative assessment of the vulnerability indices (LVI and IPCC) for female-headed and male-headed households is presented in Fig. 8.5. From Fig. 8.5 it is found that the female households are more vulnerable than male headed households based on both measures of vulnerability assessment.

Chapter 9
Adaptation Strategy of Households and Its Determinants

This chapter examines different adaptation strategies chosen by the households in order to reduce climate change vulnerability and to identify the factors responsible to adaptation across different agro-climatic regions of West Bengal.

9.1 Identifying Adaptation Strategy

We present the possible adaptation strategies undertaken by the households across five different agro-climatic regions of West Bengal and they are discussed in the following way.

First, livestock rearing is one of the adaptation strategies prevailing in the five regions. Livestock rearing includes buffalo, cow, goat, pig, hen and duck. Livestock rearing supports additional income of the households during an adverse climate situation. It can minimize the risk of climate change.

Second, the formation of self-help groups (SHGs) is one of the adaptation strategies of the households. Self help groups (SHGs) are voluntary associations of downtrodden people. It comprises with 10–15 women to form a group to solve their common socio-economic problems through mutual help. It offers loan to the households to build up their assets, increase their wealth and enable for starting small businesses and fight against risks and poverty. Through self help groups women are empowered in terms of enhancing income and decision making in their families.

Third, migration is another adaptation strategy of the households. Migration occurs due to various reasons. The hill region of Darjeeling district is famous for the tourism sector. The economy of this district is mainly based on tourists. In the non-tourist season the people of Darjeeling migrate to the nearest surrounding areas where there is an opportunity for getting jobs. In the extreme landslide situation people also migrate to other places. In the drought prone regions of Purulia district people migrate to the nearest agriculturally developed districts like Hooghly, Burdwan and Howrah

J. P. Basu, *Climate Change Vulnerability and Communities in Agro-climatic Regions of West Bengal, India*, https://doi.org/10.1007/978-3-030-50468-7_9

due to non-availability of job in the dry seasons. In coastal Sundarban and East Midnapore district, people migrate to the industrial sector like rice mill, cold storage etc. and tertiary sector (such as hotel, restaurant, shopping complex, private guard etc.) throughout the year.

Fourth, diversification of occupation is another adaptation strategy of the households. People diversify their income earning opportunity from climate dependent occupations (such as agriculture and allied activities, tea garden labour and tourist guide cum driver and fishing etc.) to non-farm activities and petty business in adverse geographical and climatic situations.

Fifth, collection of non-timber forest products (NTFPs) is the one of the adaptation strategies followed by the households. NTFPs mainly comprise fuel-wood, wild fruits and vegetables (like mushroom), medicinal plants, sal leaves for making sal dish, Kutchi kathi for stiching sal leaves, kendu pata for bidi making etc. The people collect these forest products and sell the product and earn income which supplement household's income.

Sixth, access to credit from nationalized bank is another adaptation strategy. People get loans for various purposes either for consumption or productive purposes. Productive purpose includes starting businesses, to purchase farm/non-farm implement and or to support the house repairing that are affected by disaster or climate change.

Seventh, fishing is another important adaptation strategy in the coastal areas. A major portion of households in coastal sundarban and East Midnapore district are engaged in fishing for livelihood generation. In addition, the fishing sector is the backbone of the economy in those regions. As there is no job opportunity for the people and agriculture is not lucrative due to salinity in the land, more and more people are engaged in fishing. A large number of people mainly women are also involved in shrimp-larva collection.

Lastly, crab collection is also important adaptation strategy of the people in the coastal sunderban. Most of the tribal people in coastal sunderban are dependent on crab collection for the sustenance of livelihood. The people who are landless and very poor go for the collection of crab in the river using boats. They return after 15–20 days and sell the collected amount of crab to the local arat.

9.2 Adaptation Strategies of the Households in the Hill Regions of Darjeeling District

The household in the hill regions of Darjeeling district identifies the adaptation strategies like access to bank credit, collection of non-timber forest products, formation of self-help groups, livestock rearing, migration and diversification of occupations from climate sensitive to non-climate sensitive. It is observed from Table 9.1 that about 89% of households are dependent on livestock rearing, 68% of households are dependent on the collection of non-timber forest products 51% are dependent

Table 9.1 Adaptation strategies of the households in the hill regions of Darjeeling district

Adaptation options	Number of households (%) N = 150
Access to bank credit	70 (46.67)
Collection of non-timber forest products (NTFPs)	102 (68)
Formation of self-help groups (SHGs)	77 (51.34)
Livestock rearing	133 (88.67)
Migration	30 (20)
Diversification of occupations	10 (6.67)

Source Field Survey, 2018
Note Figures in the parentheses represent percentage

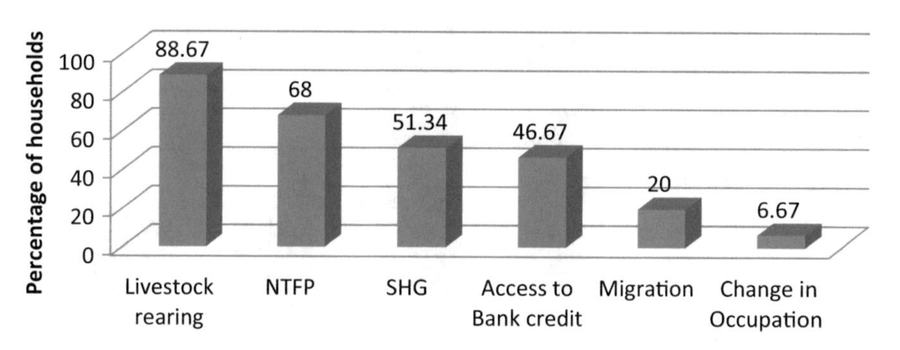

Fig. 9.1 Adaptation strategies of the households in the hill region of Darjeeling district

on self-help group and 47% of households are able to access bank credit while 20% of households are dependent on migration. Only 7% of households diversify their occupation from climate sensitive to non-climate sensitive sectors (Fig. 9.1).

9.3 Adaptation Strategies of the Households in the Foothill Region of Jalpaiguri District

The households of foothill regions of Jalpaiguri district have identified five different adaptation strategies in order to reduce vulnerability of climate change. They are access to bank credit, formation of self-help groups, livestock rearing, migration and diversification of occupations from climate sensitive to non-climate sensitive. The most important adaptation strategy in Jalpaiguri district is the formation of SHGs (see Table 9.2). 68.46% of sample households have formed SHGs. Second important

Table 9.2 Adaptation strategies of the households in foothill region of Jalpaiguri district

Adaptation options	Number of households (%)
Formation of self-help groups (SHGs)	89 (68.46)
Migration	15 (11.54)
Access to bank credit	59 (45.38)
Diversification of occupations	25 (19.23)
Livestock rearing	84 (64.62)

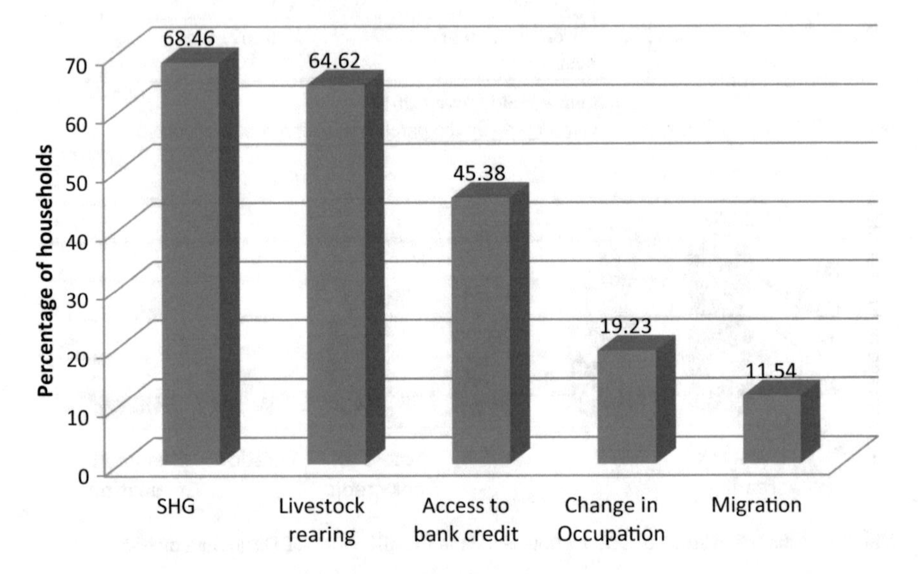

Fig. 9.2 Adaptation strategies in foothill region of Jalpaiguri district

adaptation strategy is the rearing of livestock (64.62%) which is followed by access to credit (45.38%), diversification of occupation (19.23%) and migration (11.54%) (see Fig. 9.2).

9.4 Adaptation Strategies of the Households in the Drought Regions of Purulia District

The households in the drought regions of Purulia district have identified five adaptation strategies like collection of non-timber forest products, formation of self-help groups, livestock rearing, migration and diversification of occupations from climate sensitive to non-climate sensitive. It is observed from Table 9.3 that about 78% of households are dependent on the collection of non-timber forest products while 75% of the households are dependent on livestock rearing. On the other hand, 51% of

Table 9.3 Adaptation strategies of the households in drought prone regions of Purulia district

Adaptation option	Number of households (%)
Formation of self-help groups (SHGs)	77 (51.33)
Migration	36 (24)
Diversification of occupations	23 (15.33)
Livestock rearing	113 (75.33)
Collection of non-timber forest products (NTFPs)	117 (78)

Source Field Survey, 2018
Note Figures in the parentheses represent percentage

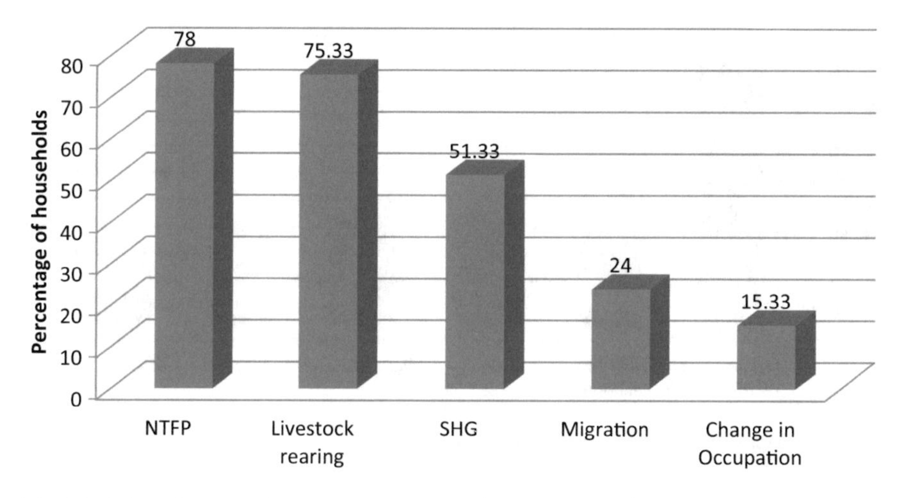

Fig. 9.3 Adaptation strategies in drought prone regions of Purulia district

households have formed self-help groups while 24% of households are dependent on migration. Only 15% of households diversify their occupation from climate sensitive to non-climate sensitive sectors (see Fig. 9.3).

9.5 Adaptation Strategies of the Households in the Coastal Region of Sundarban, South 24 Parganas District of West Bengal

The households in the coastal regions of Sunderban, South 24 Parganas district have identified six adaptation strategies like fishing, crab collection, formation of self-help groups (SHGs), livestock rearing, migration and diversification of occupation. Of these six adaptation strategies, most important adaptation strategy is livestock

Table 9.4 Adaptation strategies of the households in the coastal region of Sundarban

Adaptation options	Number of households (%)
Fishing	40 (20.31)
Crab collection	53 (26.91)
Formation of self-help groups (SHGs)	78 (39.6)
Livestock rearing	187 (94.93)
Migration	74 (37.57)
Diversification of occupations	79 (40.11)

Source Field Survey, 2018
Note Figures in the parentheses represent percentage

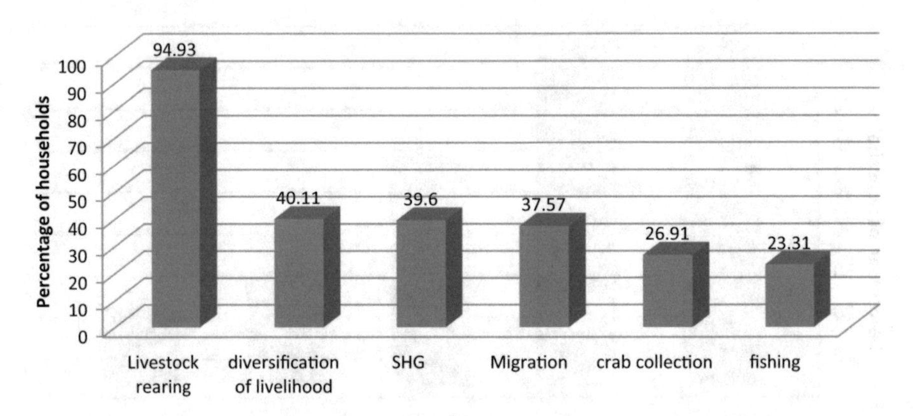

Fig. 9.4 Adaptive strategies in coastal Sundarban

rearing. It is observed from Table 9.4 that 94.93% of households are dependent on livestock rearing. The second important adaptation strategy is the diversification of occupation from climate sensitive sector to non-climate sensitive sector. It is also observed that about 40% of households are dependent on SHGs while 38% of households migrate (Table 9.4). On the other hand, 20% and 27% of households are dependent on fishing and crab collection respectively (see Fig. 9.4 and Table 9.4).

9.6 Adaptation Strategies of the Households in the Coastal Region of East Midnapore District

The households in the coastal regions of East Midnapore district have identified four adaptation strategies like the formation of self-help groups (SHGs), livestock rearing, migration and diversification of occupation. Out of these four strategies, the most important adaptation strategy is livestock rearing which is chosen by 96.23% of households as an adaptation strategy. It is observed that 59% of households have

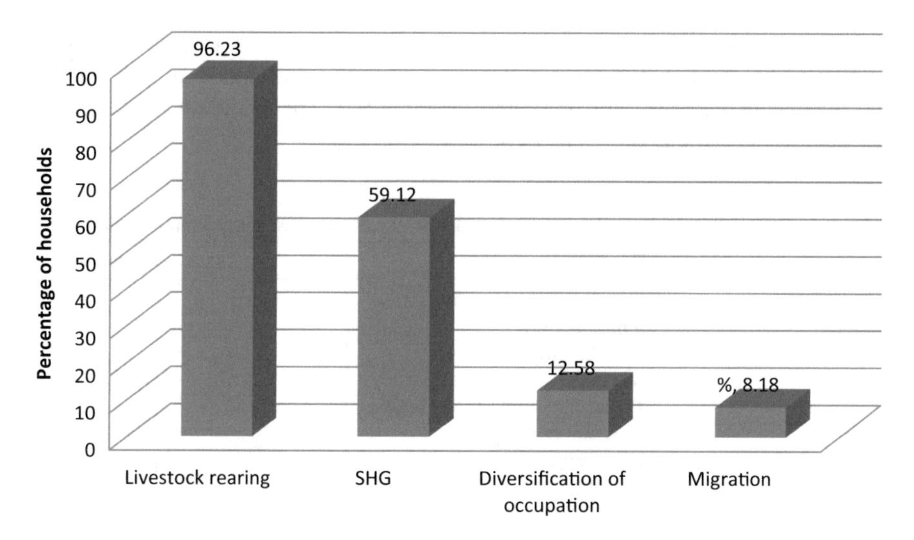

Fig. 9.5 Adaptation option in coastal region of East Midnapore district

Table 9.5 Adaptation strategies of the households in the coastal region of East Midnapore district

Adaptation options	Number of households (%)
Formation of self-help groups (SHGs)	94 (59.12)
Migration	13 (8.18)
Diversification of occupation	20 (12.58)
Livestock rearing	36 (96.23)

Source Field Survey, 2018
Note Figures in the parentheses represent percentage

formed SHGs. Here migration is very few only 8% of households are migrating. On the other hand, about 13% of households diversify their occupation from climate sensitive to non-climate sensitive sectors (see Fig. 9.5 and Table 9.5).

9.7 Factors Affecting Adaptation to Climate Change Across Agro-Climatic Regions

The objective of the present section is to identify the factors influencing adaptation strategies of the households across different agro-climatic region of West Bengal. The common adaptation strategies are livestock rearing, formation of self help groups (SHGs), migration and diversification of livelihoods and accessibility of credit. The character of each adaptation strategy is binary. If a household follows a particular adaptation strategies then it takes value "1" otherwise takes value "0". The dependent variables are adaptation strategies like livestock rearing, SHGs, migration, diversification of livelihoods and accessibility of credit. The description of independent and dependent variables are presented in Table 9.6. The basic statistics of each independent variable is shown in Table 9.7.

Table 9.6 Description of dependent and independent variables in five agro-climatic regions

	Description	Unit	Relation
Agro-climatic region dummy	D1 when hill region = 1, otherwise = 0 D2 when foot-hill = 1, otherwise = 0 D3 when drought prone region = 1, otherwise = 0 D4 when coastal region of Sundarban = 1, otherwise = 0 D5 when coastal region of East Midnapore = 1, otherwise = 0		
Household size	Total family member of the household	Person	+
Caste	SC = 1, ST = 2, OBC = 3, GEN = 4		–
Age	Age of head of the family	Years	–
Gender	Sex of the household head	Male = 1, female = 0	+
Education	Education of head of the households, years of schooling	Years	+
Marital status	Whether the head of the family married or unmarried	Married = 1 Unmarried = 0	
Male earner	Total number of adult earning male in the household	Person	+
Operational land holding	Operational land holding (= own + leased in – leased out) of cultivable land	acre	–
Access to information	Access to information through TV or mobile		+
Physical asset	Total value of the physical assets of the households	Rs	+
Off-farm income	Percentage share of total income wage labour, forestry, fishing, crab collection etc.		–
Farm income	Percentage share of total income from crops and livestocks	Rs	+
Non-farm income	Percentage share of total income from trading, transport and informal sector	Rs	–

(continued)

Table 9.6 (continued)

	Description	Unit	Relation
Climate perception index	Average of normalized score of the climate related variables like increase in rainfall, increase in hot days, increase in cold days, increase in rainy days, unusual formation of fog, increase in stormy events, increase in flood (or increase in sea level/rapid inward shift of coastline)/drought/landslide and increase in salinity of land in coastal belt etc.		+
Dependent Variables			
SHG	Whether households joined in SHGs?	Yes = 1, No = 0	
Migration	Whether households migrate or not?	Yes = 1, No = 0	
Credit accessibility	Whether the households have accessibility of institutional credit	Yes = 1, No = 0	
Diversification of livelihood	Is there any diversification from agriculture to non farm sector for livelihood?	Yes = 1, No = 0	
Livestock rearing	Are households' rear livestock?	Yes = 1, No = 0	

We have used probit model to estimate the factors affecting adaptation. The estimated results of Probit model across five agro-climatic regions are presented in Table 9.8. The estimated McFadden's Pseudo-R^2 explains the goodness of fit of the Probit model. The estimated Pseudo-R^2 s in the present model are 0.2429, 0.3353, 0.2806, 0.3888 and 0.3762 for formation of SHG, migration, access to credit, livelihood diversification and livestock rearing respectively. On the basis of the values of LR Chi-square, log likelihood and P value all the five models are overall statistically significant at 1% level.

9.8 Discussions of the Model's Result

The formation of self help groups (SHGs) is one of the statistically significant adaptation strategies to cope with variety of shocks due to climate change through raising income in different agro-climatic regions of West Bengal except in drought region of Purulia district. The formation of SHGs is influenced by age of the household's head, education, gender, marital status, physical assets and off-farm income. It is

Table 9.7 Descriptive statistics (Mean and Standard Deviation) of explanatory variables

	Drought region		Foot-hill region		Hill region		Coastal Sundarban		Coastal East Midnapore	
	Mean	Std dev.	Mean	Std dev.	Mean	Std dev.	Mean	Std dev.	Mean	Std dev.
Household size (person)	4.80	1.70	4.98	1.87	4.33	1.82	3.88	1.28	3.88	1.77
Caste SC = 1	19%	0.86	13%	1.06	–	0.74	52%	1.15	7%	0.51
ST = 2	58%		25%		73%		27%		–	
Gen = 3	12%		25%		12%		1%		93%	
OBC = 4	11%		37%		15%		20%		–	
Age (years)	43.58	10.64	38.85	13.31	48.82	16.47	43.66	10.68	45.49	13.47
Gender male%	78%	0.42	58%	0.49	50%	0.50	67%	0.47	85%	0.36
Education (years)	4.53	4.59	5.74	4.87	6.63	4.66	3.50	4.19	7.13	4.02
Marital status, married%	97%	0.16	78%	0.41	93%	0.26	99%	0.10	94%	0.24
Operational land holding (acre)	0.80	0.94	0.79	1.07	0.04	0.22	0.20	0.41	0.17	0.31
Access to information	73%	0.45	73%	0.81	100%	0.00	35%	0.48	58%	0.49
Number of male earner	1.49	0.77	0.73	0.45	1.51	0.79	1.29	0.61	1.38	0.85
Physical asset (Rs)	20,752	67,847	100,049	434,255	75,087	131,772	17,721	22,604	34,102	39,942
Farm income (% to total income)	21.33	27.61	12.54	28.46	0.52	2.59	0.90	2.46	4.86	5.13
Off-farm income (% to total income)	55.62	35.88	10.51	31.38	51.09	44.26	64.83	45.16	50.53	44.70
Non-farm income (% to total income)	23.04	32.03	62.98	44.75	48.64	44.41	34.27	44.95	44.62	45.89
Climate perception index	0.39	0.20	0.66	0.18	0.57	0.16	0.66	0.23	0.36	0.21

Source Computed by author from primary data

Table 9.8 Result of probit model across five agro-climatic regions of West Bengal

	SHG		Migration		Accessibility of credit	
	Coef.	P > |z|	Coef.	P > |z|	Coef.	P > |z|
Hill region	−0.58	0.029	−0.484	0.389	−1.827	0.002
Drought region	0.005	0.978	0.512	0.009	−0.44	0.019
Foot-hill region	0.399	0.058	−0.268	0.294	−0.248	0.239
Coastal Sundarban	0.334	0.038	0.794	0.000	0.643	0.000
Coastal East Midnapore	0.369	0.019	−0.575	0.006	0.28	0.088
Households size	−0.01	0.754	0.038	0.31	0.031	0.347
Caste	−0.013	0.802	−0.017	0.765	0.224	0.000
Age	−0.006	0.006	−0.002	0.647	0.001	0.741
Sex	−0.182	0.082	−0.166	0.171	0.202	0.059
Education	0.025	0.028	0.027	0.007	0.015	0.188
Marital status	0.225	0.055	−0.234	0.313	0.16	0.425
Operational land holding	0.014	0.852	−0.233	0.061	0.081	0.292
Access to information	0.06	0.489	0.178	0.088	−0.026	0.767
Male income earner	0.023	0.767	0.201	0.021	0.155	0.075
Physical asset	−0.017	0.083	0.001	0.863	0.024	0.008
Farm income	−0.002	0.771	−0.004	0.423	−0.008	0.099
Off-farm income	−0.007	0.058	0.01	0.033	−0.003	0.493
Non-farm income	−0.005	0.294	−0.002	0.595	0	0.959
Climate perception index	−0.078	0.731	−0.406	0.142	−0.248	0.301
No. of observation	786		786		786	
LR Chi-square	46.64		110.35		86.52	
P value	0.002		0		0	
Log likelihood	−52026		−35265		−493.54	
Pseudo R square	0.2429		0.3353		0.2806	

	Diversification of livelihood		Livestock rearing			
	Coef.	P > |z|	Coef.	P > |z|		
Hill region	−0.274	0.7	0.577	0.383	–	–
Drought region	0.642	0.008	0.843	0.000	–	–
Foot-hill region	0.424	0.11	0.976	0.000	–	–
Coastal Sundarban	1.491	0.000	0.426	0.056	–	–
Coastal East Midnapore	0.427	0.058	0.55	0.028	–	–

(continued)

Table 9.8 (continued)

	Diversification of livelihood		Livestock rearing			
	Coef.	P > lzl	Coef.	P > lzl		
Family size	0.016	0.689	0.089	0.04	–	–
Caste	−0.137	0.026	0.005	0.941	–	–
Age	−0.002	0.664	−0.002	0.744	–	–
Sex	0.029	0.824	0.031	0.817	–	–
Education	0.046	0.001	0.007	0.629	–	–
Marital status	0.113	0.656	−0.091	0.716	–	–
Operational holding	0.095	0.282	0.216	0.021	–	–
Access to information	0.24	0.068	0.002	0.987	–	–
Male income earner	0.029	0.759	−0.018	0.864	–	–
Physical asset	0	0.675	0	0.869	–	–
Farm income	−0.014	0.006	0.019	0.001	–	–
Off-farm income	−0.008	0.089	−0.014	0.004	–	–
Non-farm income	0.002	0.685	0.013	0.006	–	–
Climate Perception index	0.053	0.848	0.644	0.061	–	–
No of observation	786		786		–	–
LR Chi-square	148.38		115.88		–	–
P value	0		0		–	–
Log likelihood	−318.85		−270.99		–	–
Pseudo R square	0.3888		0.3762		–	–

observed from Table 9.8 that the formation of SHGs is significant and positively associated with education, marital status and off-farm income. On the other hand, age, gender and physical assets are inversely related to the formation of SHGs, one of the adaptation strategies of the households.

Migration is another adaptation option chosen by the households. It is found from Table 9.8 that migration is a statistically significant adaptation strategy only in the drought region and coastal belts of Indian Sunderbans and Midnapore district. Migration depends on education, operational land holdings, access to information, male income earners and off-farm income. It is observed that migration is significant and positively related to education, access to information, male income earners and off-farm income while it is negatively related to operational land holdings.

Accessibility to credit which helps to cope with climate risk is another adaptation strategy chosen by the households. It is a statistically significant adaptation option in the agro-climatic regions except foot-hill region of Jalpaiguri district of West Bengal. Accessibility to credit is influenced by caste, gender, male income earners, physical assets and farm income. Accessibility to credit is significant and positively related to caste, gender, male income earners and physical assets while on the other hand it is inversely related to farm income.

Diversification of livelihoods is also one of the adaptation strategies chosen by the households. It is significant adaptation strategy prevailing in the drought region and coastal regions of Sunderbans and Midnapore district. It depends on education, caste, access to information, off-farm income and farm income. Of these factors education and access to information are positive and significantly associated with diversification of livelihood. On the other hand, caste, off-farm income and farm income are inversely affected diversification of livelihood.

Livestock rearing is also another adaptation strategy of the households. It is a statistically significant adaptation strategy in the selected agro-climatic regions except the hill region of Darjeeling district. It depends on size of the households, operational land holdings, farm income, off-farm income, non-farm income and climate perception index. It is observed from Table 9.8 that livestock rearing is positively related to size of the household, operational land holdings, farm income and non-farm income and climate perception index. On the other hand, it is negatively related to off-farm income.

Chapter 10
Perception of Climate Change of Different Livelihood Groups of Households

The present chapter attempts to address the perceptions of climate change among livelihood groups across five agro-climatic regions of West Bengal. This chapter presents climate perception index of different livelihood groups.

10.1 Climate Change Perception of Livelihood Groups in the Hill Region of Darjeeling

In the hill region of Darjeeling district we have selected eight indicators to understand the perception of climate change of the households. The selected indicators are the decrease in rainfall, increase in rainfall, increase in landslides, increase in storm, increase in number of hot days, increase in number of cold days, increase in number of rainy days and increase in earthquake. Each response is rated in a three- point scale by yes, no and don't know. The most of occupational groups of households have expressed their response in favour of increased number of hot days, cold days, rainy days and earthquake (Table 10.1). The increase in rainfall, decrease in rainfall, increase in landslide and increase in storm, most of the households in different occupational groups have responded "No" and "Don't know" against these climate events. The perception index of each occupational groups of households is constructed by taking an average of normalized scores of the above eight perception indicators. The perception index of different occupational group of households is shown in Fig. 10.1. The perception index is highest for tourist—cum-driver (0.420) followed by petty businessmen (0.397), workers in formal sectors (0.375), casual labourer (0.350), tea garden labourer (0.348). The lowest perception index is in the workers in the informal sector (0.319).

J. P. Basu, *Climate Change Vulnerability and Communities in Agro-climatic Regions of West Bengal, India*, https://doi.org/10.1007/978-3-030-50468-7_10

Table 10.1 Climate perceptions of different livelihood groups of households in the hill region of Darjeeling district

Perception indicators		Tourist guide cum driver (No. of households = 42)	Petty business men (No. of households = 28)	Workers in informal sector (No. of households = 9)	Workers in formal sector (No. of households = 47)	Tea garden labour (No. of households = 14)	Casual labour (No. of households = 10)
Increase in rainfall	Yes	10 (23.81)	3 (10.71)	1 (11.11)	10 (21.28)	2 (14.29)	1 (10)
	No	15 (35.71)	12 (42.86)	6 (66.67)	21 (44.68)	5 (35.71)	4 (40)
	Don't know	17 (40.48)	13 (46.43)	2 (22.22)	16 (34.04)	7 (50)	5 (50)
Decrease in rainfall	Yes	10 (23.81)	5 (17.86)	1 (11.11)	9 (19.15)	1 (7.14)	1 (10)
	No	12 (28.57)	10 (35.71)	2 (22.22)	20 (42.55)	7 (50)	3 (30)
	Don't know	20 (47.62)	13 (46.43)	6 (66.67)	18 (38.3)	6 (42.86)	6 (60)
Increase in landslide	Yes	16 (38.1)	12 (42.86)	2 (22.22)	14 (29.79)	6 (42.86)	2 (20)
	No	12 (28.57)	8 (28.57)	5 (55.56)	15 (31.91)	5 (35.71)	5 (50)
	Don't know	14 (33.33)	8 (28.57)	2 (22.22)	18 (38.3)	3 (21.43)	3 (30)
Increase in storm	Yes	15 (35.71)	11 (39.29)	3 (33.33)	15 (31.91)	4 (28.57)	3 (30)
	No	12 (28.57)	8 (28.57)	5 (55.56)	14 (29.79)	4 (28.57)	4 (40)
	Don't know	15 (35.71)	9 (32.14)	1 (11.11)	18 (38.3)	6 (42.86)	3 (30)
Number of hot days increase	Yes	22 (52.38)	14 (50)	6 (66.67)	23 (48.94)	6 (42.86)	3 (30)
	No	14 (33.33)	6 (21.43)	0	8 (17.02)	3 (21.43)	3 (30)
	Don't know	6 (14.29)	8 (28.57)	3 (33.33)	16 (34.04)	5 (35.71)	4 (40)
Number of cold days increase	Yes	25 (59.52)	19 (67.86)	4 (44.44)	22 (46.81)	7 (50)	4 (40)
	No	13 (30.95)	6 (21.43)	3 (33.33)	13 (27.66)	4 (28.57)	6 (60)
	Don't know	4 (9.52)	3 (10.71)	2 (22.22)	12 (25.53)	3 (21.43)	0
Number of rainy days increase	Yes	24 (57.14)	13 (46.43)	2 (22.22)	21 (44.68)	6 (42.86)	8 (80)
	No	10 (23.81)	9 (32.14)	5 (55.56)	13 (27.66)	4 (28.57)	1 (10)
	Don't know	8 (19.05)	6 (21.43)	2 (22.22)	13 (27.66)	4 (28.57)	1 (10)
Increase in earthquake	Yes	19 (45.24)	12 (42.86)	4 (44.44)	27 (57.45)	7 (50)	6 (60)
	No	15 (35.71)	11 (39.29)	1 (11.11)	9 (19.15)	7 (50)	3 (30)
	Don't know	8 (19.05)	5 (17.86)	4 (44.44)	11 (23.4)	0	1 (10)

Source Field Survey, 2018
Note Figures in the parentheses represent percentage

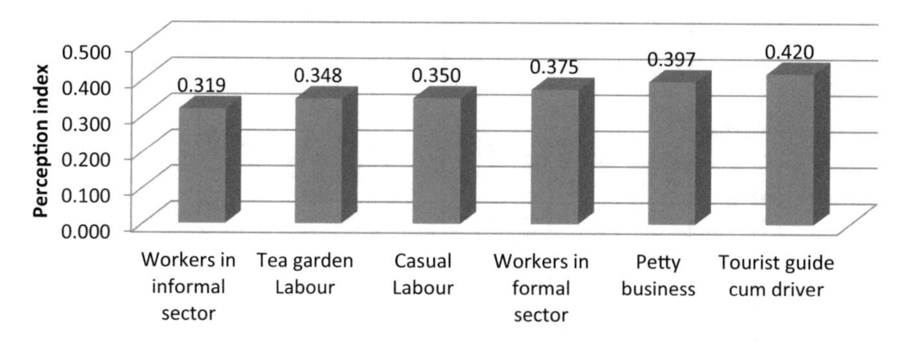

Fig. 10.1 Perception index of different occupational groups of households

10.2 Climate Change Perception of Livelihood Groups in the Foothill Region of Jalpaiguri

In the foot-hill region the study has chosen five climate indicators such as decrease in rainfall, increase in rainfall, increase in the number of hot days, increase in the number of cold days and increase in the number of rainy days. The perception of different livelihood groups on climate variables is shown in Table 10.2. It is observed that more than 50% cultivators have expressed their responses in favour of increase in rainfall, the increase in the number of hot days and the increase in number of cold days. The forest dependent communities responded in favour of decrease in rainfall, the increase in the number of hot days and the number of increase in cold days. The other livelihood groups of households also expressed their responses in favour of the number of hot days increased.

The perception indices of different occupational groups of households are shown in Fig. 10.2. The perception index is highest for cultivators (0.547) followed by casual labour (0.541), forest dependent communities (0.533), workers in the formal sector (0.523). The lowest perception index is found to be the workers who are in the informal sector (0.472).

10.3 Climate Change Perception of Livelihood Groups in Drought Region of Purulia District

In the drought region of Purulia district we have has chosen seven indicators of perception of climate change. These indicators are the decrease in rainfall, increase in rainfall, increase in number of hot days, increase in number of cold days, increase in storm, increase in drought and increase in number of rainy days. The perception of different livelihood groups of households is shown in Table 10.3. It is found from this table that more than 70% of cultivators and more than 50% of forest dependent

Table 10.2 Climate perception of different livelihood groups in the foothill region of Jalpaiguri district

Perception indicators		Cultivators (No. of households = 19)	Forest Dependent communities (No. of households = 6)	Workers in informal Sector (No. of households = 75)	Workers in formal Sector (No. of households = 13)	Casual Labour (No. of households = 17)
Decrease in rainfall	Yes	4 (21.05)	5 (83.33)	28 (37.33)	–	6 (35.29)
	No	6 (31.58)	–	21 (28)	7 (53.85)	7 (41.18)
	Don't know	9 (47.37)	1 (16.67)	26 (34.67)	6 (46.15)	4 (23.53)
Increase in rainfall	Yes	10 (52.63)	1 (16.67)	37 (49.33)	12 (92.31)	9 (52.94)
	No	5 (26.32)	3 (50)	14 (18.67)	–	1 (5.88)
	Don't know	4 (21.05)	2 (33.33)	24 (32)	1 (7.69)	7 (41.18)
Number of hot days increased	Yes	16 (84.21)	4 (66.67)	43 (57.33)	10 (76.92)	12 (70.59)
	No	2 (10.53)	–	23 (30.67)	3 (23.08)	–
	Don't know	1 (5.26)	2 (33.33)	9 (12)	–	5 (29.41)
Number of cold days increased	Yes	18 (94.74)	6 (100)	57 (76)	12 (92.31)	15 (88.24)
	No	1 (5.26)	–	11 (14.67)	–	1 (5.88)
	Don't know	–	–	7 (9.33)	1 (7.69)	1 (5.88)
Number of rainy days increased	Yes	4 (21.05)	–	12 (16)	–	4 (23.53)
	No	7 (36.84)	3 (50)	44 (58.67)	8 (61.54)	10 (58.82)
	Don't know	8 (42.11)	3 (50)	19 (25.33)	5 (38.46)	3 (17.65)

Source Field Survey, 2018
Note Figures in the parentheses represent percentage

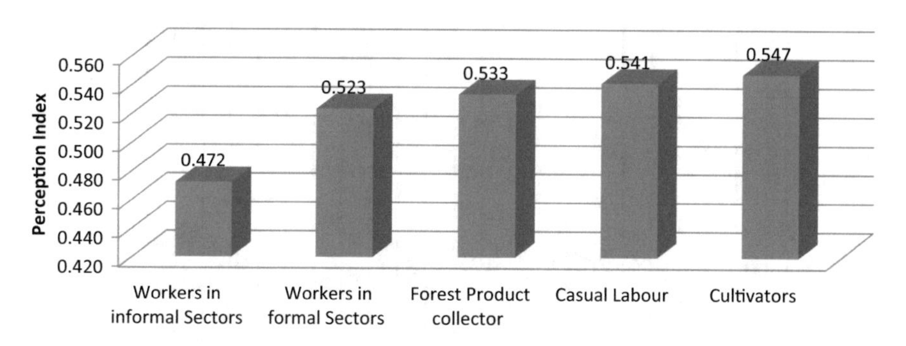

Fig. 10.2 Perception index of livelihood groups in foot-hill region of Jalpaiguri district

communities have realized only the increased in storm. They do not have more perception of other climatic events.

The perception indices of different livelihood groups are shown in Fig. 10.3. The perception index is highest for workers in informal sector (0.384) followed by cultivators (0.359), forest dependent communities (0.322) and casual labourer (0.243) (Fig. 9.3).

10.4 Climate Change Perception of Livelihood Groups of Households in Coastal Regions of Sundarban in South 24 Parganas

In coastal region of Sundarban we have chosen twelve indicators of the perception of climate change. These indicators are the decrease in rainfall, increase in rainfall, increase in sea level, increase in fish stock in the river, increase in number of hot days, increase in number of cold days, increase in number of rainy days, unusual formation of fog, increase in storm, rapid/more inward shift of coastline, permanent encroachment of new areas by saline water and increase in salinity level in land. The perception of different livelihood groups is presented in Table 10.4. It is observed from this table that more than 50% of the petty businessmen have realized the increase in salinity in land. More than 50% of workers in the informal sector, petty business men and casual labourers have felt the increased in seal level.

It is also observed that among the livelihood groups perception index is maximum for workers in petty businessmen (0.380) followed by fishing communities (0.360), workers in informal sector (0.332), casual labourer (0.327) and crab collecting households (0.322) (Fig. 10.4).

Table 10.3 Climate perception of different livelihood groups in drought prone Purulia district

Perception indicators		Cultivators (No. of households = 49)	Forest dependent communities (No. of households = 39)	Workers in informal sectors (No. of households = 42)	Casual Labour (No. of households = 20)
Decrease in rainfall	Yes	22 (44.9)	15 (38.46)	22 (52.38)	11 (55)
	No	16 (32.65)	10 (25.64)	6 (14.29)	4 (20)
	Don't know	11 (22.45)	14 (35.9)	14 (33.33)	5 (25)
Increase in rainfall	Yes	10 (20.41)	5 (12.82)	13 (30.95)	1 (5)
	No	22 (44.9)	18 (46.15)	16 (38.1)	11 (55)
	Don't know	17 (34.69)	16 (41.03)	13 (30.95)	8 (40)
Increase in storm	Yes	36 (73.47)	25 (64.1)	28 (66.67)	9 (45)
	No	6 (12.24)	10 (25.64)	9 (21.43)	7 (35)
	Don't know	7 (14.29)	1 (2.56)	5 (11.9)	3 (15)
Increase in drought	Yes	7 (14.29)	7 (17.95)	6 (14.29)	1 (5)
	No	10 (20.41)	13 (33.33)	5 (11.9)	9 (45)
	Don't know	32 (65.31)	19 (48.72)	31 (73.81)	10 (50)
Number of hot days increased	Yes	27 (55.1)	16 (41.03)	25 (59.52)	8 (40)
	No	17 (34.69)	14 (35.9)	10 (23.81)	10 (50)
	Don't know	5 (10.2)	9 (23.08)	7 (16.67)	2 (10)
Number of cold days increased	Yes	14 (28.57)	11 (28.21)	13 (30.95)	2 (10)
	No	23 (46.94)	21 (53.85)	18 (42.86)	11 (55)
	Don't know	12 (24.49)	7 (17.95)	11 (26.19)	7 (35)
Number of rainy days increased	Yes	7 (14.29)	7 (17.95)	6 (14.29)	2 (10)
	No	29 (59.18)	21 (53.85)	22 (52.38)	11 (55)
	Don't know	13 (26.53)	11 (28.21)	14 (33.33)	7 (35)

Source Field Survey, 2018
Note Figures in the parentheses represent percentage

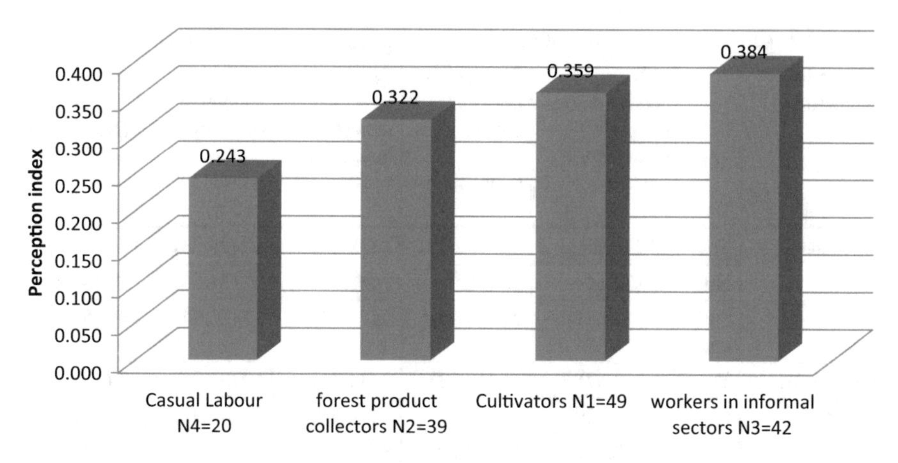

Fig. 10.3 Perception index by livelihood groups in drought prone Purulia district

10.5 Climate Change Perception of Livelihood Groups of Households in Coastal Region of East Midnapore

In coastal region of East Midnapore we have chosen eleven indicators of climate perception. These indicators are the decrease in rainfall, increase in rainfall, increase in salinity level in land, increase in sea level, increase in fish stock in the river, increase in number of hot days, increase in number of cold days, increase in number of rainy days, unusual formation of fog, increase in storm and more inward shift of coastline.

Climate perception of different livelihood groups in Coastal region of East Midnapore district is presented in Table 10.5. It is found from this table that most of the households in different livelihood groups have less perception of climate change.

It observed that perception index is highest for casual labourer (0.372) followed by workers in formal sector (0.356), van puller (0.338), workers in informal sector (0.333), and cultivators households (0.317) and fishing households (0.298) (Fig. 10.5).

Table 10.4 Climate perception of different livelihood groups in coastal Sundarban

Perception indicators		Fishing communities (No. of households = 28)	Workers in informal sectors (No. of households = 61)	Petty business men (No. of households = 18)	Crab collecting communities (No. of households = 52)	Casual labourer (No. of households = 38)
Decrease in rainfall	Yes	9 (32.14)	12 (19.67)	7 (38.89)	10 (19.23)	6 (15.79)
	No	9 (32.14)	28 (45.9)	7 (38.89)	27 (51.92)	13 (34.21)
	Don't know	10 (35.71)	21 (34.43)	4 (22.22)	15 (28.85)	19 (50)
Increase in rainfall	Yes	8 (28.57)	11 (18.03)	2 (11.11)	12 (23.08)	2 (5.26)
	No	11 (39.29)	24 (39.34)	8 (44.44)	22 (42.31)	20 (52.63)
	Don't know	9 (32.14)	26 (42.62)	8 (44.44)	18 (34.62)	16 (42.11)
Increase in salinity of land	Yes	13 (46.43)	30 (49.18)	10 (55.56)	14 (26.92)	13 (34.21)
	No	8 (28.57)	15 (24.59)	3 (16.67)	24 (46.15)	12 (31.58)
	Don't know	7 (25)	16 (26.23)	5 (27.78)	14 (26.92)	13 (34.21)
Increase in sea level	Yes	7 (25)	34 (55.74)	11 (61.11)	16 (30.77)	19 (50)
	No	10 (35.71)	11 (18.03)	2 (11.11)	20 (38.46)	10 (26.32)
	Don't know	11 (39.29)	16 (26.23)	5 (27.78)	16 (30.77)	9 (23.68)
Increase in fish stock	Yes	11 (39.29)	28 (45.9)	13 (72.22)	14 (26.92)	16 (42.11)
	No	12 (42.86)	17 (27.87)	4 (22.22)	17 (32.69)	10 (26.32)
	Don't know	5 (17.86)	16 (26.23)	1 (5.56)	21 (40.38)	12 (31.58)
Increase in hot days	Yes	5 (17.86)	19 (31.15)	2 (11.11)	21 (40.38)	14 (36.84)
	No	14 (50)	21 (34.43)	12 (66.67)	19 (36.54)	9 (23.68)
	Don't know	9 (32.14)	21 (34.43)	4 (22.22)	12 (23.08)	15 (39.47)
Increase in cold days	Yes	10 (35.71)	14 (22.95)	6 (33.33)	20 (38.46)	15 (39.47)
	No	8 (28.57)	25 (40.98)	9 (50)	20 (38.46)	11 (28.95)
	Don't know	10 (35.71)	22 (36.07)	3 (16.67)	12 (23.08)	12 (31.58)
Increase in rainy days	Yes	11 (39.29)	21 (34.43)	4 (22.22)	14 (26.92)	9 (23.68)
	No	9 (32.14)	18 (29.51)	7 (38.89)	16 (30.77)	11 (28.95)
	Don't know	8 (28.57)	22 (36.07)	7 (38.89)	22 (42.31)	18 (47.37)
Increase in unusual formation of fog	Yes	10 (35.71)	18 (29.51)	6 (33.33)	14 (26.92)	16 (42.11)
	No	10 (35.71)	23 (37.7)	8 (44.44)	16 (30.77)	10 (26.32)

(continued)

Table 10.4 (continued)

Perception indicators		Fishing communities (No. of households = 28)	Workers in informal sectors (No. of households = 61)	Petty business men (No. of households = 18)	Crab collecting communities (No. of households = 52)	Casual labourer (No. of households = 38)
	Don't know	8 (28.57)	20 (32.79)	4 (22.22)	22 (42.31)	12 (31.58)
Increase in storm	Yes	11 (39.29)	17 (27.87)	9 (50)	21 (40.38)	11 (28.95)
	No	11 (39.29)	20 (32.79)	5 (27.78)	15 (28.85)	9 (23.68)
	Don't know	6 (21.43)	24 (39.34)	4 (22.22)	16 (30.77)	18 (47.37)
Rapid/more inward shift of coastline	Yes	12 (42.86)	21 (34.43)	8 (44.44)	22 (42.31)	14 (36.84)
	No	9 (32.14)	21 (34.43)	7 (38.89)	18 (34.62)	12 (31.58)
	Don't know	7 (25)	19 (31.15)	3 (16.67)	12 (23.08)	12 (31.58)
Permanent encroachment of new areas by saline water	Yes	14 (50)	18 (29.51)	4 (22.22)	23 (44.23)	14 (36.84)
	No	8 (28.57)	23 (37.7)	10 (55.56)	20 (38.46)	15 (39.47)
	Don't know	6 (21.43)	20 (32.79)	4 (22.22)	9 (17.31)	9 (23.68)

Source Field Survey, 2018
Note Figures in the parentheses represent percentage

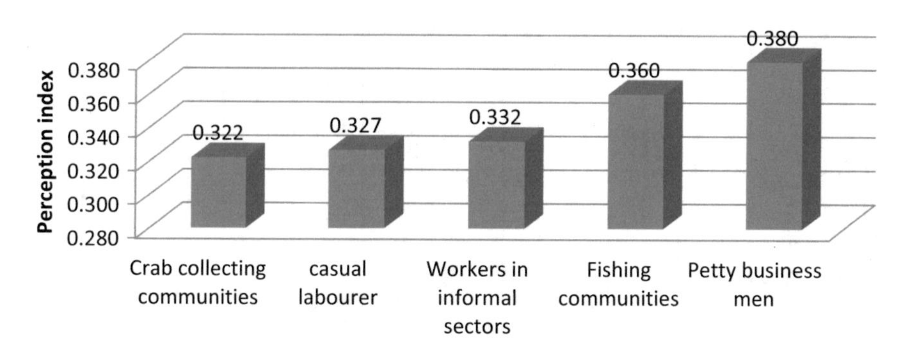

Fig. 10.4 Perception index of different livelihood groups in coastal Sundarban

Table 10.5 Climate perception of livelihood groups in coastal regions of East Midnapore

Perception indicators		Casual labour (No. of households = 28)	Cultivators (No. of households = 20)	Fishing communities (No. of households = 26)	Workers in informal sector (No. of households = 21)	Worker in formal sector (No. of households = 26)	Van Puller (No. of households = 38)
Decrease in rainfall	Yes	9 (32.14)	8 (40)	8 (30.77)	9 (42.86)	13 (50)	13 (34.21)
	No	9 (32.14)	3 (15)	6 (23.08)	9 (42.86)	7 (26.92)	14 (36.84)
	Don't know	10 (35.71)	9 (45)	12 (46.15)	3 (14.29)	6 (23.08)	11 (28.95)
Increase in rainfall	Yes	14 (50)	7 (35)	8 (30.77)	9 (42.86)	10 (38.46)	15 (39.47)
	No	8 (28.57)	4 (20)	6 (23.08)	5 (23.81)	9 (34.62)	9 (23.68)
	Don't know	6 (21.43)	9 (45)	12 (46.15)	7 (33.33)	7 (26.92)	14 (36.84)
Increase in salinity of land	Yes	9 (32.14)	6 (30)	10 (38.46)	6 (28.57)	11 (42.31)	15 (39.47)
	No	7 (25)	9 (45)	9 (34.62)	8 (38.1)	6 (23.08)	7 (18.42)
	Don't know	12 (42.86)	5 (25)	7 (26.92)	7 (33.33)	9 (34.62)	16 (42.11)
Increase in sea level	Yes	13 (46.43)	5 (25)	7 (26.92)	6 (28.57)	8 (30.77)	9 (23.68)
	No	7 (25)	8 (40)	14 (53.85)	9 (42.86)	9 (34.62)	14 (36.84)
	Don't know	8 (28.57)	7 (35)	5 (19.23)	6 (28.57)	9 (34.62)	15 (39.47)
Increase in fish stock	Yes	13 (46.43)	7 (35)	9 (34.62)	8 (38.1)	11 (42.31)	11 (28.95)
	No	7 (25)	7 (35)	11 (42.31)	10 (47.62)	11 (42.31)	11 (28.95)
	Don't know	8 (28.57)	6 (30)	6 (23.08)	3 (14.29)	4 (15.38)	16 (42.11)
Increase in hot days	Yes	6 (21.43)	9 (45)	10 (38.46)	8 (38.1)	13 (50)	15 (39.47)
	No	10 (35.71)	4 (20)	5 (19.23)	7 (33.33)	4 (15.38)	11 (28.95)
	Don't know	12 (42.86)	7 (35)	11 (42.31)	6 (28.57)	9 (34.62)	12 (31.58)
Increase in cold days	Yes	7 (25)	7 (35)	3 (11.54)	4 (19.05)	11 (42.31)	13 (34.21)

(continued)

Table 10.5 (continued)

Perception indicators		Casual labour (No. of households = 28)	Cultivators (No. of households = 20)	Fishing communities (No. of households = 26)	Workers in informal sector (No. of households = 21)	Worker in formal sector (No. of households = 26)	Van Puller (No. of households = 38)
	No	9 (32.14)	5 (25)	12 (46.15)	11 (52.38)	7 (26.92)	10 (26.32)
	Don't know	12 (42.86)	8 (40)	11 (42.31)	6 (28.57)	8 (30.77)	15 (39.47)
Increase in rainy days	Yes	10 (35.71)	5 (25)	9 (34.62)	8 (38.1)	10 (38.46)	10 (26.32)
	No	6 (21.43)	12 (60)	6 (23.08)	5 (23.81)	7 (26.92)	12 (31.58)
	Don't know	12 (42.86)	3 (15)	11 (42.31)	8 (38.1)	9 (34.62)	16 (42.11)
Increase in unusual formation of fog	Yes	13 (46.43)	7 (35)	9 (34.62)	6 (28.57)	8 (30.77)	14 (36.84)
	No	7 (25)	6 (30)	7 (26.92)	8 (38.1)	8 (30.77)	15 (39.47)
	Don't know	8 (28.57)	7 (35)	10 (38.46)	7 (33.33)	10 (38.46)	9 (23.68)
Increase in storm	Yes	12 (42.86)	4 (20)	7 (26.92)	3 (14.29)	7 (26.92)	17 (44.74)
	No	9 (32.14)	10 (50)	10 (38.46)	5 (23.81)	8 (30.77)	8 (21.05)
	Don't know	7 (25)	6 (30)	9 (34.62)	13 (61.9)	11 (42.31)	13 (34.21)
Rapid/more inward shift of coastline	Yes	9 (32.14)	3 (15)	4 (15.38)	9 (42.86)	5 (19.23)	13 (34.21)
	No	10 (35.71)	4 (20)	7 (26.92)	6 (28.57)	10 (38.46)	9 (23.68)
	Don't know	9 (32.14)	13 (65)	15 (57.69)	6 (28.57)	11 (42.31)	16 (42.11)
Permanent encroachment of new areas by saline water	Yes	10 (35.71)	8 (40)	9 (34.62)	8 (38.1)	4 (15.38)	9 (23.68)
	No	10 (35.71)	7 (35)	8 (30.77)	9 (42.86)	12 (46.15)	15 (39.47)
	Don't know	8 (28.57)	5 (25)	9 (34.62)	4 (19.05)	10 (38.46)	14 (36.84)

Source Field Survey, 2018
Note Figures in the parentheses represent percentage

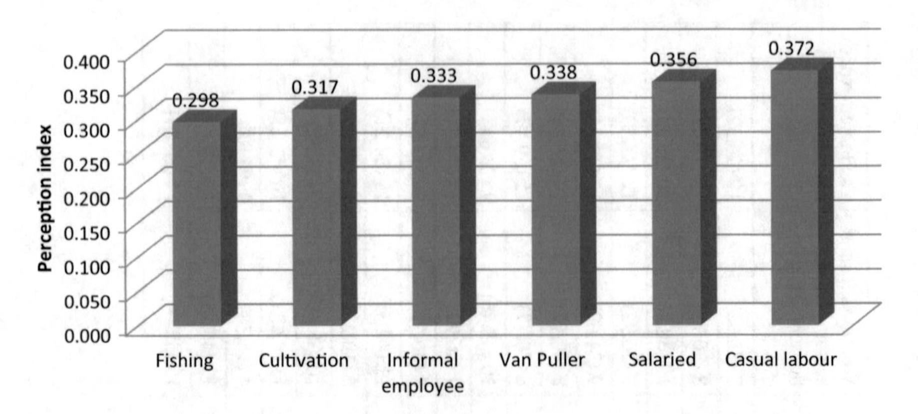

Fig. 10.5 Perception index of livelihood groups in Coastal region of East Midnapore

Chapter 11
Conclusions and Policy Recommendations

This chapter concludes the findings of the study, governmental policies to reduce vulnerability, identifies key vulnerabilities and recommendations.

11.1 Conclusions

The following conclusions have emerged from the above the study;

First, the socio-economic conditions of the households are weak across five agro-climatic regions of West Bengal. Most of the households comprise landless, marginal and small landholdings, scheduled tribes and scheduled caste. These five agro-climatic regions are dominated by poor people as more than 50% of people are under BPL (Below Poverty Line). There have been high illiteracy rates in the foot-hill region, drought prone region and coastal region of Sundarban. The majority of households are living in kachha houses.

Second, most of the households are not accessing infrastructural facilities like facilities for drinking water, sanitation, irrigation and borrowing facilities from banks. The major modes of transportation in the study areas are bus, bicycle and van rickshaw. The important livestock assets owned by the households are hen, duck, goat and pig etc.

Third, major sources of livelihood of the households in the district of Darjeeling are Government services (employment in the government sector like police and army), tourist guide cum driving and tea garden. The major sources of livelihoods of the households in the drought region of Purulia district are agriculture, informal employment, forest product collection and casual work. Fishing, crab collection, and informal employment are the major sources of livelihoods in the coastal sundarbans. The major sources of livelihoods in the foot-hill region of Jalpaiguri district are informal employment, cultivation, casual labour, employment in the formal sector, and forest product collection. There are six major sources of livelihoods like casual

J. P. Basu, *Climate Change Vulnerability and Communities in Agro-climatic Regions of West Bengal, India*, https://doi.org/10.1007/978-3-030-50468-7_11

labour, cultivation, fishing, jobs in informal sectors, and van pulling business in the coastal region of East Midnapore.district.

Fourth, the monthly consumption expenditure per household differs substantially across different agro-climatic regions of West Bengal. In the hill regions of Darjeeling district the monthly consumption expenditure per household is Rs. 7558, in the foothill regions of Jalpaiguri district it is Rs. 824; in the drought regions of Purulia district it is Rs. 726.68; in coastal Sundarban it is Rs. 2332.38; and in coastal regions of East Midnapore district it is Rs. 4285.60. This shows that the people in Darjeeling and East Midnapore districts are better off than other regions of West Bengal in terms of consumption per household.

Fifth, the vulnerability indices of 18 districts of West Bengal based on secondary data shows that the district South 24 Parganas was the most vulnerable while the district Nadia was least vulnerable among 18 districts of West Bengal in 2001. After a decade the order of vulnerability has changed in 2011 and the district Purulia reaches the most vulnerable district and East Midnapore becomes least vulnerable district. The study also reveals that number of more vulnerable districts has risen in 2011 compared to that of 2001. The determinant of vulnerability based on secondary data has been determined using panel regression model. The determinants of vulnerability are identified as maximum temperature, cropping intensity, main agricultural population, net sown area and poverty. The maximum temperature is positive and significantly related to the vulnerability. This means that the maximum temperature has the impact on vulnerability. It is also noted that the cropping intensity and the main agricultural population are positive and significantly related to vulnerability and the net sown area is positive and significantly associated with vulnerability. It is also evident from the analysis that poverty is positive and significantly related to vulnerability. This reflects that poverty has an impact on vulnerability. That is, as poverty increases the vulnerability also increases and vice versa.

Sixth, the degree of vulnerability of the households is measured by livelihood vulnerability index. We have chosen weighted Livelihood Vulnerability Index (LVI) of Hahn et al. (2009) and modified LVI_ IPCC index to measure such climate change vulnerability. The overall weighted LVI of the households in the hill region of Darjeeling district is found to be 0.5866 while the vulnerability in terms of modified LVI_IPCC is 0.5881. The households are classified into less vulnerable, moderate vulnerable and highly vulnerable on the basis of the values of Livelihood vulnerability indices (LVI). It is found that 46% of households fall in the moderate vulnerable while 44.67% of households belong to highly vulnerable in the hill region of Darjeeling district. This indicates that there is a predominance of moderate and highly vulnerable households in the hill region of the Darjeeling district of West Bengal. The overall weighted LVI and modified LVI_IPCC of the households in the foot-hill region of Jalpaiguri district are found to be 0.5505 and 0.5343 respectively. It is found that the 36% of households fall in the moderate vulnerable while 33% of households belong to highly vulnerable and 31% of households belong to less vulnerable categories. This shows that there is a predominance of moderate and highly vulnerable households in the foot- hill region of Jalpaiguri district of West Bengal. On the other hand, the overall weighted LVI of the households in the drought region of Purulia district is found to be 0.6076 while in terms of modified LVI_IPCC the vulnerability index is 0.6029. In the drought region of Purulia district about 57% of the sample households belong to highly vulnerable, 29% of households belong to moderate vulnerability and only 13% households belongs to less vulnerable. This shows that most of the household are highly vulnerable. The overall weighted LVI of households in the coastal sundarban is found to be 0.5980 while the modified LVI_IPCC index is 0.5843. In the coastal region of Sundarban majority of the households (55.33%) are highly vulnerable, 23% are moderately vulnerable while 22% belong to less vulnerable. The overall weighted LVI and modified LVI_IPCC of households in the coastal region of East Midnapore district are found to be 0.4471 and 0.3978 respectively. It is found that the majority of the households (55%) are moderate vulnerable while 21% belong to highly vulnerable and 24% of households belong to less vulnerable categories.

On the basis of the livelihood vulnerability index the degree of vulnerability of the households in the drought prone district of Purulia is found to be highest on account of lack of employment opportunity, persistence of high illiteracy, lack of irrigation facility and mono-cropping cultivation. On the other hand, the degree of vulnerability of the households in the hill regions of Darjeeling district and coastal regions of East Midnapore district is relatively lower compared to other regions of West Bengal because of the fact that these two districts are of tourist oriented regions of West Bengal. Thus, the tourism industry plays an important role in explaining lower vulnerability.

Seventh, to order to identify the factors responsible for the vulnerability of the households we have carried out order logit model of pooled data over five agro-climatic regions Using this model, the factors affecting vulnerabilities are identified like number of adaptation strategies, caste, gender, education, land holdings, access to climate information, off-farm income and non-farm income. Number of adaptation strategies and vulnerability are negatively related. If more and more of adaptation strategies are taken the level of vulnerability will be reduced and vice versa. There is a negative relationship between caste and vulnerability. This implies that lower caste households (such as SC and ST) are more vulnerable compared to higher caste (OBC and General) households. Gender is inversely related to vulnerability. It means that female headed households are more vulnerable than male headed households. Education is negatively related to vulnerability. It reveals that higher the education level of households head lower will be the vulnerability of households. Land holding is inversely related to the vulnerability. Accessibility of climatic information and vulnerability is inversely related. This means that higher the accessibility to climate information there will be lower vulnerability. Non-farm income and vulnerability is also inversely related. It implies that higher the percentage share of non-farm income to total income lower will be the vulnerability of those households. Off-farm incomes and vulnerability is positively related. It implies that higher the percentage share of off-farm income to total income higher will be the vulnerability of the households.

Eight, in the hill region of Darjeeling district tea garden workers are the most vulnerable group of households while formally employed workers are less vulnerable. In the districts of Jalpaiguri and Purulia the most vulnerable group of households are casual labourer and forest dependent community respectively. The least vulnerable group of households in the Purulia and Jlpaiguri districts is workers in the formal sector. Crab collecting communities and fishing communities are the most vulnerable group of households whereas petty businessmen are the least vulnerable group in the coastal Sundarban of West Bengal. Casual labourer and fishing communities are the most vulnerable groups in the coastal East Midnapore district of West Bengal and workers in the informal sector are the least vulnerable group.

Ninth, the livelihood vulnerability indices for female and male headed households are found to be 0.4556 and 0.4483 respectively in the hill regions of Darjeeling district. The modified livelihood indices for female and male are found to be 0.5109 and 0.5057 respectively. Both measure (LVI and modified LVI_IPCC) shows that female headed households are more vulnerable than male headed households in Darjeeling district. The livelihood vulnerability indices for female and male-headed households are found to be 0.4090 and 0.4175 in the foot-hill region of Jalpaiguri district respectively. The modified livelihood vulnerability indices (LVI_IPCC) for female and male headed households are 0.4384 and 0.4701 respectively in the district of Jalpaiguri. This shows that the male headed households are more vulnerable than female headed households in the foot-hill region of Jalpaiguri district.

In the drought region of Purulia district the vulnerability indices for female and male headed households are observed to as 0.5909 and 0.5237 respectively. In terms of modified (LVI-IPCC) the vulnerability indices for female households and male headed households are found to be 0.572 and 0.4912 respectively. Both the findings showed that female headed households are more vulnerable than male headed households in the drought region of Purulia district.

The livelihood vulnerability indices for female and male-headed households are found to be 0.5600 and 0.4246 respectively in the coastal region of Sundarbans. In terms of modified vulnerability indices for female and male headed households are found to be 0.5621 and 0.4763 respectively. This means that the female headed households are more vulnerable than male headed households.

In the coastal region of East Midnapore district the livelihood vulnerability indices for female and male headed households are found to be 0.5420 and 0.463 respectively. In terms of modified LVI_IPCC the overall vulnerability indices for female and male headed households are found to be 0.5450 and 0.4673 respectively. On the basis of these values the result showed that the female households are more vulnerable than male headed households in the district of East Midnapore.

Tenth, the study has identified eight adaptation strategies in the selected five agro-climatic regions of West Bengal. They are access to credit from banks, collection of non-timber forest product (NTFP), formation of Self Help Group (SHG), livestock rearing, migration, diversification of occupation from climate sensitive sector like agriculture to non-climate sensitive sector like non farm sector, petty business etc.

11.2 Key Vulnerabilities

The present study has identified the key vulnerabilities of the households across the five agro-climatic regions of West Bengal. The key vulnerabilities identified in the hill region of Darjeeling district are scarcity of water, landslides and storm. In the foot-hill region of Jalpaiguri district the key vulnerabilities are insufficiency of water for drinking as well as irrigation purposes, low landholdings, low per capita income and low diversification of livelihood opportunities. In the drought region of Purulia district the key vulnerabilities are lack of education, scarcity of water for irrigation and cultivation, scarcity of water for drinking purposes and low opportunity for income generation activities and employment creation. In the coastal region of Sundarban the key vulnerabilities are flood due to sea level rise, cyclone, low per capita income, lack of education, lack of health facilities, lack of employment opportunity and lack of income generating activities. In the coastal region of East Midnapore district the key vulnerabilities are low land holdings, less availability of job opportunity, high dependency on female earners, and lack of early warning information on extreme climate events.

On the basis of the above results it is observed that the first hypothesis has been accepted. That means the casual labourers, workers in the informal sectors, tea garden labourers, cultivators are more vulnerable than petty businessmen and workers in the formal sector. The second hypothesis is rejected in the sense that the drought region of Purulia district is most vulnerable district compared to other agro-climatic regions of West Bengal. The results also establish that the third hypothesis is accepted. The vulnerability of the households is influenced by non-climatic factors vis-à-vis climatic factors. That means the socio-economic factors also influencing the climate change vulnerability.

11.3 Government Policy on Vulnerability and Adaptation

The present study has an important policy implication for the reduction of vulnerability and poverty and maintaining a sustainable livelihood security of the vulnerable people in different agro-climatic regions of West Bengal. The goal of adaptation is to cope with climate risk by empowering the poor and vulnerable people for a better standard of living in the long run. In the circumstances, India launched National Adaptation Policy for Climate Change (NAPCC) in 2008 as a signatory of UNFCCC to address the issues of climate change like adaptation and mitigation with the aim of ensuring sustainable development and high economic growth rates. The national adaptation policy covers national mission for sustainable agriculture, national mission for green India, national mission for sustaining the Himalayan ecosystem, national water mission and national mission on strategic knowledge for climate change. Later four national missions were included in 2014 viz. Wind Energy, Health, Coastal Areas and Waste to Energy (Dey et al. 2016).

In view of the above guidelines of NAPCC, every State of India directed to formulate State Action Plan on Climate Change to address climate change concern. The State of West Bengal launched its Climate Change Action Plan in 2011 and 2012. Their plan and policies on climate change adaptation incorporated the climate related sectors like Disaster Management, Agriculture, Water Resources, Forestry, Coastal Zone Management, Rural Development, Fisheries, Health, Energy, Rural Electrification, Poverty Alleviation, and Women Empowerment in the River Delta.

West Bengal State Action Plan on Climate Change (WBSAPCC 2012) reported that the traditional farmers used indigenous varieties of seeds of agricultural crops which are climate tolerant and fight against climate change. At the same time, the farmers followed Integrated Farming System with the combination of crops, fisheries and livestock to ensure self-sustainability and alternative livelihoods.

In addition, West Bengal gave an important priority on the construction of embankments and dykes under Flood Control/Management activities and took initiatives for raising irrigation coverage, encouragement of rain water harvesting for portable water and construction of portable tank water to avoid contamination.

The State has taken various initiatives for expanding the Crop Insurance packages for small and marginal farmer's security against crop loss during flood or cyclone in the state (WBSAPCC 2012). The state has also arranged Early Warning System (EWS) in the coastal Sundarban to combat the stress of cyclones and storm surges under Disaster Risk Reduction (WBSAPCC 2012). The plantation and regeneration of mangrove forests on the degraded mud flats are on the top priority in Coastal Sundarban. This gives rise to ensure natural protection of island from cyclone and storms. There are several programs of the central and state governments working in coastal Sundarban like National Rural water and Sanitation Program, National Elementary Education Program (Sarva Shiksa Abhiyan), Mahatma Gandhi National Rural Employment Guarantee Act (MGNREGA), the housing scheme, Indira Awas Yojana, the Food for Work Programme, and the rural road building scheme, Pradhan Mantri Grameen Sadak Yojana. These programs are important for rural development and vulnerability reduction measures.

11.4 Recommendations

Given the above policy and programs of the government of India, the general recommendations are made in connection with the reduction of climate change vulnerability.

First, is to identify high and moderate vulnerable households in different agro-climatic regions to strengthen the adaptive capacity of those households.

Second, since poverty is chronic in rural West Bengal the policy interventions in the agro-climatic regions should focus on assisting poor households to accumulate assets through increased investment and employment generation that enhances their mean consumption level.

Third, the reduction of consumption variability is through either reducing exposure to climate risk or improving the ex post coping mechanisms of the vulnerable households. In this context, capacity building measures of households like efforts to improve income, provision of education, accessibility to health care, safe drinking water, employment opportunity; irrigation facilities etc. are the major policy suggestions to reduce vulnerability.

Fourth, measures should be taken to strengthen the livelihood resiliency of the households. In order to do that the Governmental supportive measures such as providing shelter to the homeless through Indira Awas Yojana/Pradhan Mantri Awas Yojana, more working days in Mahatma Gandhi National Rural Employment Guarantee Act (MGNREGA), Promoting productive activity through formation of Self Help Groups (SHGs), providing regular safe drinking water supply, irrigation facility, Pradhan Mantri Fasal Bima Yojna.

Fifth, women should be given more priority to empower and to access opportunities for income generating activities so that they can diversify their livelihoods to reduce their vulnerability.

Sixth, the tourism sector plays an important role for the reduction of vulnerability. That is, emphasis should be given on the development of eco-tourism to reduce vulnerability.

Seventh, education is one of the key factors in building the resilience level of households to climate change impacts. Therefore, priority should be given on the development of education of the head of the households to reduce vulnerability.

After the identification of key vulnerabilities in different agro-climatic regions, the specific recommendations are made in those regions of West Bengal in the context of vulnerability reduction.

First, in the hill regions of Darjeeling district the recommendations are as follows.

(1) In Darjeeling district, emphasis should be given to construct more water reservoir to the storage of rain water and spring water in order to provide regular water supply to the people.
(2) To check landslide importance should be given on controlling deforestation as well as illegal construction in hill.

Second, in the foot-hill regions of Jalpaiguri district the recommendations are as follows.

(1) Extension of irrigation is urgently needed in Jalpaiguri district to develop multiple cropping systems.
(2) For steady increase in income of households importance should be given on the diversification of livelihoods.

Third, in the drought regions of Purulia district the recommendations are as follows.

(1) Most lacking area in this region is education. Priority should be given to primary level education.
(2) Importance should be given on rain water harvesting by excavating and maintaining of water bodies like ponds and tanks through MGNREGA that attracts more employment opportunities and helps to raise production and productivity of crops.
(3) Aforestation in the degraded or barren land (such as fruit garden) through MGNREGA/SHG that create income and employment opportunity.
(4) Emphasis should be given on the proper implementation of development policies of the government (such as housing scheme, MGNREGA, old age pension, distribution of seed and other farming equipment) so that the fruits of such policies should reach to the targeted down trodden people.
(5) Setting up agri-business industries (like tomato juice, sugar industry, wine industry from mahua fruits, bamboo basket and toy industry, biri Industry from kendu pata) is much urgent to provide more income and employment opportunities.

Fourth, in the coastal regions of Sundarban, in the district of South 24 Parganas, the recommendations are as follows.

(1) Government should emphasize social sector development like health, education and sanitation etc. and development of infrastructures like road, bridges, boats, jetties, electricity, cyclone shelters and cyclone resistance buildings etc.
(2) Building up seawall to protect people from sea level rise.
(3) Arrangement of early warning system to inform the people regarding cyclone and storm.
(4) Plantation of more Mangrove forests to protect the people from cyclone and storms etc.
(5) Importance should be given on alternative sources of livelihood such as shrimp and crab collection.

Fifth, in the coastal regions of East Midnapore district, the recommendations are as follows.

(1) The government of West Bengal should give priority on the development of the tourism sector in Digha, Mondarmoni, Sankarpur and Tajpur in the East Midnapore district.
(2) Development of handicraft industries like mat, ornaments of Jhinuk (oyster), bamboo (toys) to increase employment and income generation for women.
(3) Arrangement of early warning system regarding cyclone and storm such that information of such extreme climate events should reach to the people in that region.

References

Dey S, Ghosh AK, Hazra S (2016) Review of West Bengal State adaptation policies, Indian Bengal delta. In: Deltas, vulnerability and climate change: migration and adaptation [DECCMA] Working Paper. DECCMA Consortium, Southampton, UK. https://generic.wordpress.soton.ac.uk/deccma/resources/working-papers/. Last accessed 6 Aug 2018

Hahn MB, Riederer AM, Foster SO (2009) The livelihood vulnerability index: a pragmatic approach to assessing risks from climate variability and change—a case study in Mozambique. Glob Environ Change 19(1):74–88. https://doi.org/10.1016/j.gloenvcha.2008.11.002

WBSAPCC (2012) West Bengal state action plan on climate change (WBSAPCC). Department of Environment, Government of West Bengal

Printed in the United States
by Baker & Taylor Publisher Services